AF364431

Computational Techniques for Power Systems Analysis

Computational Techniques for Power Systems Analysis

Dipu Sarkar
Asst. Professor
Dept. of Electrical & Electronics Engg,
National Institute Technology Nagaland, India.

Chandan Kumar Chanda
Professor,
Dept. of Electrical Engg,
Indian Institute of Engineering Science & Technology,
Shibpore, Howrah, WB, India.

Abhinandan De
Assoc. Professor
Dept. of Electrical Engg,
Indian Institute of Engineering Science & Technology,
Shibpore, Howrah, WB, India.

BSP **BS Publications**
An Imprint of **BSP Books Pvt. Ltd.**
4-4-309/316, Giriraj Lane,
Sultan Bazar, Hyderabad - 500 095.

Computational Techniques for Power Systems Analysis
by **Dipu Sarkar, Chandan Kumar Chanda, Abhinandan De**

© 2024, *by Publisher,* All rights reserved.

Published by:

BS Publications

An Imprint of **BSP Books Pvt. Ltd.**
4-4-309/316, Giriraj Lane,
Sultan Bazar, Hyderabad - 500 095.
Phone :040 - 23445688
e-mail :info@bspbooks.net
www.bspbooks.net

ISBN: 978-93-91910-83-9 (Hardback)

Dedication

This book is dedicated to my father Late Sri Ranjit Sarkar, my mother Late Rita Sarkar, my beloved wife Bharati Sarkar and my son Suvam Sarkar.

Dipu Sarkar

This book is dedicated to my parents and my family members.

Abhinandan De

This book is dedicated to my parents and my family members and two grandsons Greg and Ayan.

Chandan Kumar Chanda

Preface

A well established and an exciting field, Electrical Engineering broadly involves working with power system with its large size, complex and integrated nature. The book has given strong emphasis on the basic principles of power systems analysis to fulfil the needs of under-graduate and post-graduate students, research-scholars as well as practicing power system engineers. In this book large scale system analysis follows a natural extension of the basic principles and presented in such a manner that all under-graduate and post graduate students can easily grasp the subject and highly appreciate them. A novel feature of this book is the inclusion of the MATAB based Power System Analysis Toolbox (PSAT), which can be practically useful for the analysis of many power system problems.

Key features of the book:

- The topics have been presented in a graded manner.
- Presentation of the classical methods of system analysis.
- Clearly illustrated worked out examples.
- Step by step problem solving techniques, which should enable the students to sharpen their problem-solving skills.

This book should be useful for the UG, PG students of all Engineering Colleges and Technical Universities and for those who will be appearing for competitive examinations like GATE, AMIE, IAS, IES, different state and central level recruitment examinations, UPSC, PSC, Railway board etc.

The authors are greatly indebted to all the staff and the management of the publishing house.

The authors also appreciate the patience and inspiration provided by their family members during the entire journey.

Constructive criticism and/or comments about the book are always welcome.

-Authors

Contents

Preface ....(v)

Chapter 1: Per Unit Representation

1.0   Introduction   1

Exercise   **64**

Chapter 2: Introduction to Fault Analysis

2.0   Introduction   67

2.1   Transients due to Balanced Fault   67

2.2   Transients owing to a 3 – Phase Alternator
Terminal Short Circuit   71

Exercise   **72**

Chapter 3: Symmetrical Fault Analysis

3.1   Necessity for Short Circuit Calculation   73

3.2   Symmetrical Fault Calculation   73

3.3   Application of Thevenin Theorem   73

      3.3.1   Symmetrical Short Circuit Analysis   73

3.4   Generator Reactors   83

3.5   Feeder Reactor   84

3.6   Bus Bar Reactor   84

Exercise   **85**

Answers   **85**

Chapter 4: The Analytical Concept for Symmetrical Components

4.1   Introduction   86

      4.1.1   Concept of 'a' Operator   86

4.2   Sequence Component to Unbalance Phase Conversion   87

4.3   Unbalance Phase to Sequence Component Conversion   88

4.4   Sequence Impedances and Sequence Networks   96

Exercise   **104**

(x)　*Contents*

Chapter 5: Asymmetrical Faults in Power Systems

5.1	Introduction	107
5.2	Single Line-To-Ground Fault	107
5.3	Line-To-Line Fault	111
5.4	Double Line-to-Ground Fault	112
Problems		114
Exercise		**115**
Answers		**116**

Chapter 6: Load Flow Analysis

6.1	Introduction	117
6.2	Classification of Bus	117
6.3	Power Flow Equation	118
6.4	Admittance Bus Matrix	118
6.5	The Algorithm for the Gauss-Seidel Iteration	126
	6.5.1　Importance of Gauss Seidel Method in Power Flow Analysis	128
	6.5.2　Analysis	128
6.6	Newton-Raphson Method of Load Flow Analysis	168
	6.6.1　Analytical Concept of N-R Load Flow Technique	169
	6.6.2　Load Flow Solution with Newton–Raphson (NR) Method	170

Chapter 7: Economic Load Dispatch

7.1	Introduction of Economic Load Dispatch	193
7.2	Economic Scheduling	193
7.3	Formulation of Economic Load Dispatch Problem	194
	7.3.1　Constraints	194
	7.3.2　The Problem Objective Function	194
	7.3.2.1　The Fuel Cost Function of Power Plant	194
Exercise		**221**
Answers		**222**

Chapter 8: Power Angle Stability and its Analysis

8.1   Introduction   224

8.2   Definitions   224

8.3   Types of Stability   224

8.4   Power Angle Equation   225

Solved Problem   **227**

8.5   Techniques for getting better Steady State Stability   228

8.6   Swing Equation   228

8.7   Analytical Concept of Swing Equation   229

8.8   Swing Curves   230

    8.8.1   Use of Swing Curves   230

8.9   Constants used in Stability Analysis   231

    8.9.1   Inertia Constant M   231

    8.9.2   Kinetic Energy N   231

    8.9.3   Inertia Constant H   231

    8.9.4   Equivalent H Constant   232

    8.9.5   Equivalent M Constant of Two Machines   233

    8.9.6   Equal Area Criterion of Stability   234

8.10 Critical Clearing Angle   240

Exercise   **246**

Chapter 1

Per Unit Representation

1.0 Introduction

The per unit value of any quantity is defined as the ratio of its actual value to another arbitrarily selected value of quantity of same dimensions. It is convenient to express impedances, voltages, powers and currents in terms of p.u. in the process of computation in power systems.

$$\text{Per unit value } \Delta = \frac{\text{Actual}}{\text{Base}}$$

i.e ratio of actual value to the base value of the same dimension.

$$\text{For any quantity X, } X_{p.u} = \frac{X}{X_{Base}}$$

Dividing the numerical value by a chosen base value of the same dimensions, any quantity can be converted to a per unit quantity. Percent quantities and per unit quantities are different by a factor of 100. Per unit quantities are dimensionless.

Suppose, a base voltage of 10 KV is considered, voltages of 10, 15, 20 KV can be shown as, 1.00, 1.50, 2.00 per unit respectively. In terms of percentage, 100, 150, 200 percent.

Let us consider,

I_A = Actual current, Amperes

I_B = Base current, Amperes

V_A = Actual voltage, Volts

V_B = Base voltage, Volts

Z_Ω = Actual impedance, Ohms

Z_b = Base impedance, Ohms

S_{va} = Actual Volt ampere

S_b = Base Volt ampere

Per unit current, $I_{pu} = I_A/I_B$

Per unit voltage, $V_{pu} = V_A/V_B$

Per unit impedance, $Z_{pu} = Z_\Omega/Z_B$

$Z_\Omega = R_\Omega + j\ X_{pu}$; $Z_{pu} = Z_\Omega/Z_b = R_\Omega/Z_B = jX_\Omega/Z_B$

or $Z_{pu} = R_{pu} + jX_{pu}$

$R_{pu} = R_\Omega/Z_B$ and $X_{pu} = X_\Omega/Z_B$

Per unit volt-ampere, $S_{pu}\ \Delta = S_{va}/S_B$

$S = P + jQ = VI\ \times$; $S_{pu} = S_{VA}/S_B = P/S_B + jQ/S_B$

or $S_{pu} = P_{pu} + jQ_{pu}$

$P_{pu} = P_{watt}/S_B$ and $Q_{pu} = Q_{watt}/S_B$

For the single phase circuit,

$Z_B = V_B/I_B$ and $S_B = V_B I_B$

and,

Base current $I_B = S_B/V_B$

$$= \frac{\text{Base KVA}}{\text{Base KV}}$$

Formulae for one base to another base:

$$Z_{pu.2} = Z_{pu.1} \times (S_{b2}/S_{b1}) \times (V_{b1}/V_{b2})^2$$

Example 1: A single phase Transformer is rated 110/440V, 4 KVA having Leakage reactance at LT side is 0.2 Ω, Determine the Leakage reactance in p.u.

Solution: Given: $V_B = 110V$, $S_B = 4$ KVA, $Z_{Act} = 0.2\Omega$

We know that,

$$Z_B = \frac{(\text{BaseKV})^2}{\text{BaseMVA}}$$

$$Z_B \text{ at LT side} = \frac{\left(110 \times 10^{-3}\right)^2 \times 1000}{4} = 3.025\ \Omega$$

$$Z_{p.u.} = \frac{z_{Act}}{z_{Base}} = \frac{0.2}{3.025} = 0.0661 \text{ p.u.}$$

For Secondary side:

$$Z_1' = Z_1\left(\frac{N_2}{N_1}\right)^2 \quad \text{where, } N_1 \& N_2 \text{ are the no. of turns on Primary and Secondary side.}$$

$$Z_1' = 0.2\left(\frac{440}{110}\right)^2 = 3.2$$

and $\qquad Z_B = \dfrac{\left(440 \times 10^{-3}\right)^2 \times 1000}{4} = 48.4\,\Omega$

$$Z_{p.u} = \dfrac{3.2}{48.4} = 0.0661 \text{ p.u.}$$

Example 2: The following figure represent a part of electrical network with component. Determine the p.u. reactance of the components:

Synchronous Generator: 25MVA, 11KV, $X'' = 0.15$ p.u.

Transformer T_1: 40MVA, 12.5Δ/132Y KV, X =0.1 p.u

Line: $200 + j500\,\Omega$

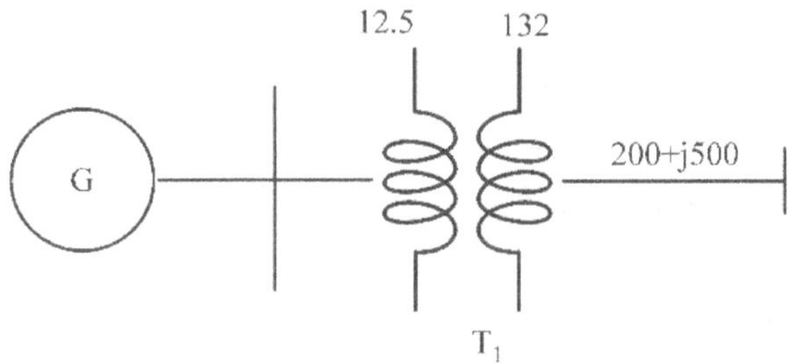

Fig.1.1

Draw the impedance diagram for the system shown in Fig.1.1. For Transmission line choose the base value as 132KV and 25MVA.

Solution: For Generator:

The Base uV of transmission line is 132 uV. Hence, base KV at the Generator side 12.5 KV

$$X'' = j0.15 \times \frac{25}{25} \times \left(\frac{11}{12.5}\right)^2$$

$$= j0.116 \text{ p.u}$$

For **Transformer 1:**

Base KV at the transformer (T_1) secondary side is 132 uV

$$X'' = j0.1 \times \frac{25}{40} \times \left(\frac{132}{132}\right)^2$$

$$= j0.0625 \text{p.u}$$

For **Transformer 2:**

$$X'' = j0.1 \times \frac{25}{30} \times \left(\frac{132}{132}\right)^2$$

$$= j0.083 \text{p.u.}$$

For Motor:

$$X'' = j0.15 \times \frac{25}{20} \times \left(\frac{11}{11}\right)^2$$

$$= j0.187 \text{ p.u.}$$

For Transmission Line:

$$Z_B = \frac{(BaseKV)^2}{BaseMVA}$$

$$= \frac{(132)^2}{25} = 696.96\Omega$$

$$Zp.u = \frac{Z_{Act}}{Z_{Base}} = \frac{200 + j500}{696.96} = 0.286 + j0.717$$

For a Load of 5MVA and 0.8 p.f lagging,

P = S cosØ = 5 × 0.8 = 4 MW

Q = S sinØ = 5 × 0.6 = 3 MVAr

If load is represented as series impedance,

$$R_{p.u.} = V_{p.u}^2 . S_B . \frac{P}{P^2 + Q^2} \quad \text{and}$$

$$X_{p.u.} = V_{p.u}^2 . S_B . \frac{Q}{P^2 + Q^2}$$

Here, $V_{p.u} = \dfrac{11}{11} = 1 \text{ p.u}$

$S_B = 25$ MVA

So,
$$R_{p.u} = (1)^2 \times 25 \times \frac{4}{4^2 + 3^2} = 4 \text{ p.u}$$

$$X_{p.u} = (1)^2 \times 25 \times \frac{3}{4^2 + 3^2} = 3 \text{ p.u}$$

P.U Impedance Diagram:

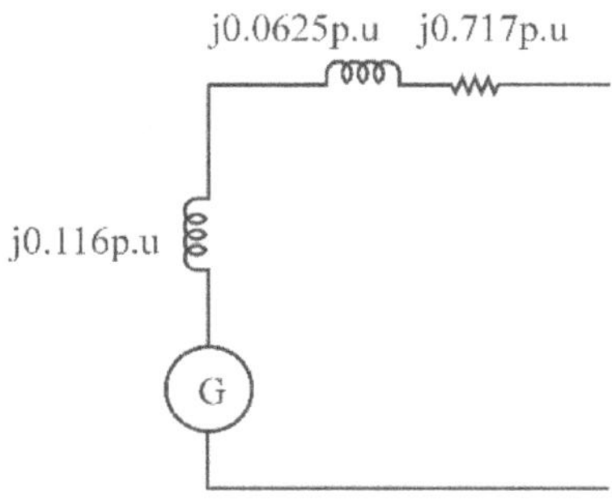

Fig.1.2

Example 3: A 3 bus system is given in Fig.1.3. The ratings of the various components are listed below:

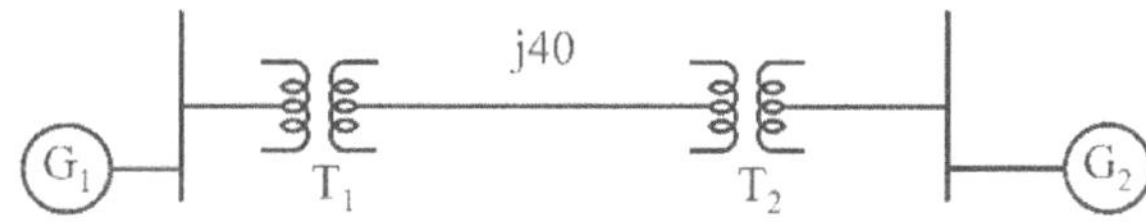

Fig.1.3

Generator 1:40MVA, 12.8KV, $X'' = 0.2$ p.u

Generator 2:50MVA, 12.2KV, $X'' = 0.15$ p.u

Transformer1:40MVA, 12KV/100KV Y/Y,X = 0.1p.u

Transformer2:20MVA, 13.5KV/125KV Y/Y,X = 0.20p.u

The line impedance are shown in Fig.1.3. Determine the reactance diagram based on 40 MVA and 12.8 KV as base quantities in generator 1.

Solution: Base MVA = 40; Base KV = 12.8

Generator 1:

$$Z_{\text{p.u.new}} = Z_{\text{p.u.given}} \times \left(\frac{\text{Base KV given}}{\text{Base KV new}} \right)^2 \times \left(\frac{\text{Base MVA new}}{\text{Base MVA given}} \right)$$

Per Unit Reactance = 0.2 × (12.8/12.8)² × (40/40)

$$= 0.0.2$$

Transmission Line 1:

Base voltage along the transmission line whose impedance is j40Ω = 12.8 × 100/12 =106kV

$$\text{Base impedance} = \frac{(\text{Base KV})^2}{\text{Base MVA}}$$

$$= 106^2/40 = 280.9 \ \Omega$$

Per unit impedance = 40/280.9

$$= 0.1423 \text{p.u}$$

Generator 2:

Base voltage = $106 \times 12.5/10.5$

$$= 12 \text{ KV}$$

$$X_{G2} = 0.15 \times (12.2/12)^2 \times 40/50$$

$$= 4.4652 \text{p.u}$$

Transformer 1:

40MVA, 12KV, X = 0.1p.u

Base values; 40MVA, 12.8KV

$$X_{T1} = 0.1 \times (12/12.8)^2 \times 40/40$$

$$= 0.0878 \text{p.u}$$

Transformer 2:

20MVA, 11.5KV, X = 0.20p.u

Base Voltage = $106 \times 13.5/125$

$$= 11 \text{KV}$$

$$X_{T2} = 0.1 \times (11.5/11)^2 \times 40/20$$

$$= 0.2185 \text{p.u}$$

Transformer 3:

Base Voltage on the H.T side of transformer 3 is 10 KV

$$X_{T3} = 0.2 \times (115/108)^2 \times 40/35$$

$$= 0.25916 \text{p.u}$$

Generator 3:

Base voltage on the bus = $106 \times 13.5/125$

$$= 11 \text{ KV}$$

20 MVA, 12KV, $x'' = 0.20$ p.u

Base values: 40MVA, 11KV

$$X_{G3} = 0.20 \times (12/11)^2 \times 40/20$$

$$= 0.476 \text{p.u}$$

Transmission line 2:

Base KV = 106

Base MVA = 40

$$\text{Base impedance} = \frac{106^2}{40}$$

$$= 280.9 \ \Omega$$

Per unit impedance = j20/280.9

$$= j0.0711$$

Reactance Diagram:

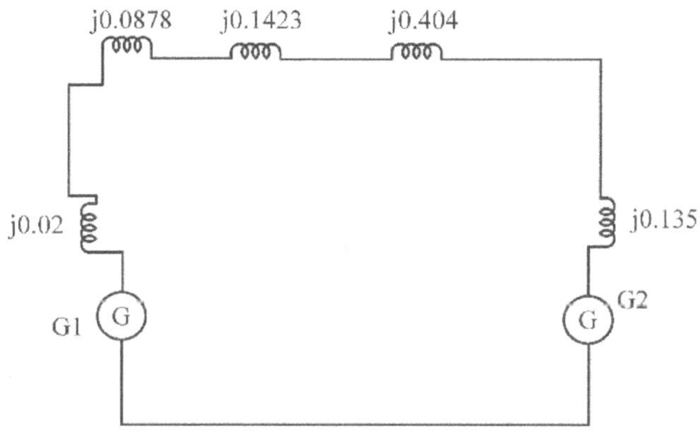

Fig.1.4

Example 4: Draw the impedance diagram for the power system shown in Fig.1.Use a Base of 45KVA,20KVat generator G1. The ratings of the generators and transformers are:

Generator 1: 20MVA, 18KV, $X'' = 20\%$,

Generator 2: 20MVA, 20KV, $X'' = 20\%$,

Synchronous motor 3, 35 MVA, 20KV, $X'' = 20\%$,

Transformer 1: Three phase Y-Y Transformers: 20 MVA, 148 Y/20 Y KV, X = 10%,

Transformer 2: Three phase Y-Δ Transformers: 10MVA,148Y/20ΔKV, X = 10%,

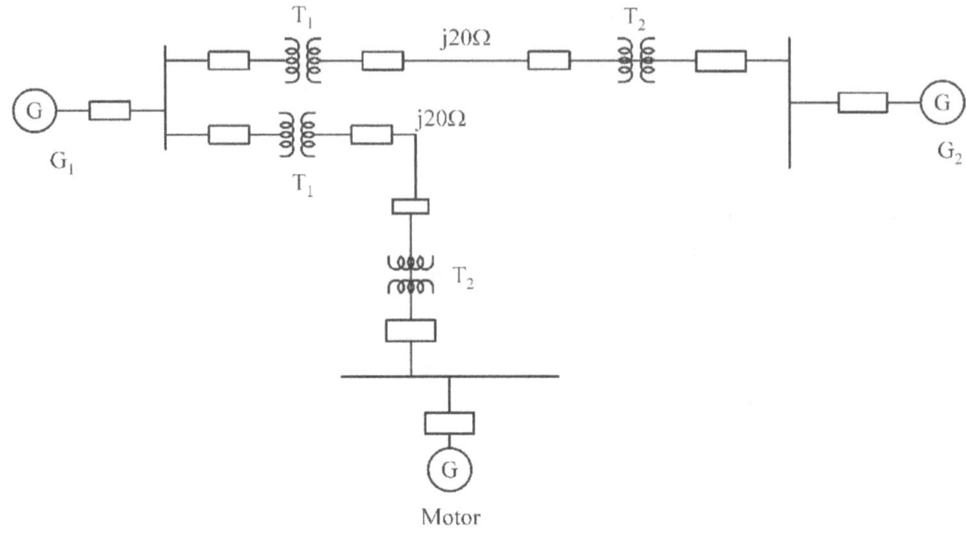

Fig.1.5

Solution: Generator 1:

Given: 25MVA, 18KV, $x'' = 20\%$

Base values: 45KVA, Base KV = 20

$$Z_{p.unew} = Z_{p.ugiven} \times \left(\frac{\text{Base KV given}}{\text{Base KV new}} \right)^2 \times \left(\frac{\text{Base MVA new}}{\text{Base MVA given}} \right)$$

$X_{G1} = 0.2 \times (18/20)^2 \times (45/20 \times 10^3)$

$\qquad = = 0.2916$ p.u.

Transformer 1:

Given: 25MVA, 20KV, X = 10%

Base values: 45KVA, 20KV

$X_{T1} = 0.1 \times (45/20 \times 10^3)$

$\qquad = 0.00025$p.u

Transmission Lines:

Base impedance = (Base KV)²/Base MVA

Base impedance = $148^2/45 \times 1000$

$\qquad\qquad = 547600\Omega$

Per unit impedance = Actual impedance/Base impedance

p.u impedance = 0.0000821p.u

p.u impedance of the other transmission line = 20/547600 = 0.0000807p.u

Generator 2:

$X_{G2} = X_{G1} = 0.0003645$p.u

Y-Δ transformer

Given: 10MVA, 128Y/12.8kΔKV, X = 10%,

Base values; 45KVA, 128KV

$X = (0.1 \times 45/10 \times 10^3)$

$\quad = 0.0005$p.u

Motor:

Given: 35MVA, 12.8KV, $x'' = 20\%$

Base values: 45KVA, 128KV

$X_m = 0.2 \times 45/35 \times 10^3$

$\quad = 0.0000257$p.u

Impedance diagram:

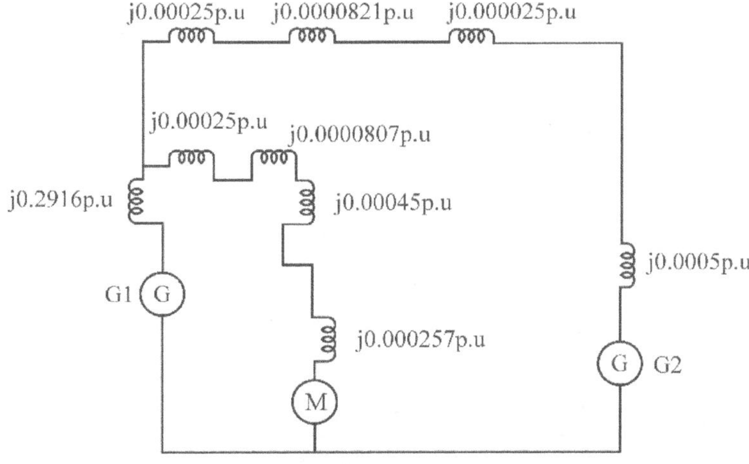

Fig.1.6

Example 5: Three generators are rated as follows:

Generator 1 – 120MVA, 40KV, reactance 14%.

Generator 2 – 170 MVA, 35KV, reactance 9%.

Generator 3 – 140MVA, 31KV, reactance 16%.

Determine the reactance of the generator corresponding to base values of 250MVA and 45 KV.

Solution:

Generator 1:

120MVA, 40kV, $x'' = 0.14$

$$Z_{p.u.new} = Z_{p.u.given} \times \left(\frac{\text{Base KV given}}{\text{Base KV new}}\right)^2 \times \left(\frac{\text{Base MVA new}}{\text{Base MVA given}}\right)$$

$$X'' = 0.14 \times (40/45)^2 \times 250/120$$

$$= 0.230 \text{ p.u.}$$

Generator 2:

170MVA, 35KV, $X'' = 0.09$

$$X'' = 0.09 \times (35/45)^2 \times 250/170 = 0.080 \text{ p.u.}$$

Generator 3:

140MVA, 31KV, $X'' = 0.16$

$$X'' = 0.16 \times (31/45)^2 \times (250/140)$$

$$= 0.1350$$

Example 6: Determine the generator voltage for the given system shown in Fig. 1.7.

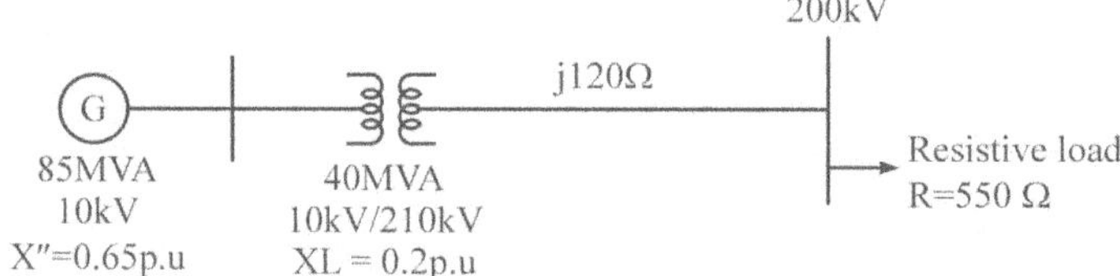

Fig.1.7

Solution: Assume a base of 120MVA and 200KV in the transmission line

Generator:

$$Z_{p.u.new} = Z_{p.u} \times \left(\frac{\text{Base KV given}}{\text{Base KV new}}\right)^2 \times \left(\frac{\text{Base MVA new}}{\text{Base MVA given}}\right)$$

$$x_G'' = 0.65 \times (10/9.523)^2 \times (120/85)$$

$$= 1.0118 \text{ p.u.}$$

Transformer:

$$X_T = 0.2 \times (210/200)^2 \times (120/40)$$

$$= 0.6615$$

Transmission lines:

Base impedance $= (\text{BaseKV})^2/\text{Base MVA}$

Base impedance = $(200)^2/120$

$\qquad$ = 333 Ω

Per unit impedance = Actual impedance/Base impedance

P.U impedance = 120/333 = 0.3603 p.u.

Resistive load of 550Ω

$R_{p.u}$= 550/333 = 1.6516 p.u.

Impedance diagram:

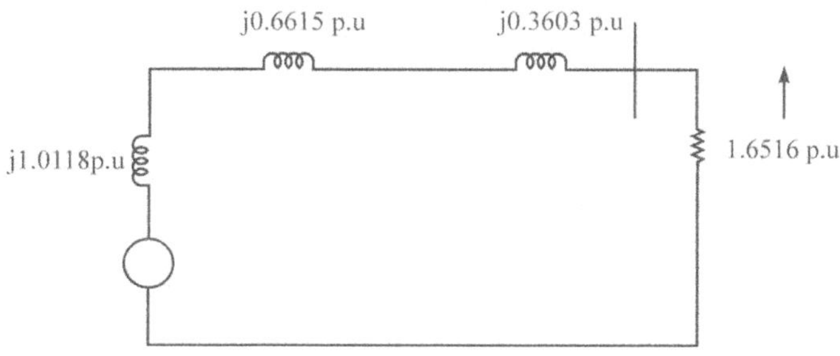

Fig.1.8

At the load bus:

$\quad$ $V_{p.u}$ = 1 p.u.

$\quad$ $R_{p.u}$ = 1.6516 p.u.

$\quad$ $I_{p.u}$ = $V_{p.u.}$

$\qquad$ = 1/1.6516 = 0.605 p.u.

Let $I_{p.u.}$ be taken as the reference phasor.

$\quad$ $I_{p.u.}$ = 0.605 + j0 = 0.605<0°

By Kirchhoff's voltage law,

Voltage drop in the network = $I_{p.u}$ [$R_{p.u}$ + j (X_G + X_T + X_L)]

$\qquad$ = 0.605 [1.6516 + j (1.0118 + 0.6615 + 0.3603)]

$\qquad$ = 0.605 [1.6516 + j2.0336]

$\qquad$ = 0.605 × 2.6197<50.9159

$\qquad$ = 1.589 p.u.

Actual generator terminal voltage:

$\quad$ V_G = $V_{p.u}$ × Base voltage at the generator terminals

$\qquad$ = 1.5849 × 9.523

$\qquad$ = 15.0930 KV

Example 7: A 200 MVA, 66 KV Three phase generator has a subtransient reactance of 20%. The generator is connected to 3 motors through a transmission line and two transformers. The motors have rated input of 40 MVA, 30 MVA and 45 MVA at 33 KV with 15% subtransient reactance. The 3phase transformer are rated at 220 MVA, 64/220 KV 'Y' with leakage reactance of 10%. The line has a reactance of 45 Ohms. Selecting the generator rating as the quantities in generator circuit, determine the base quantities in other parts of the system and evaluate the corresponding per unit values.

Solution:

Transmission line diagram for given statement is:

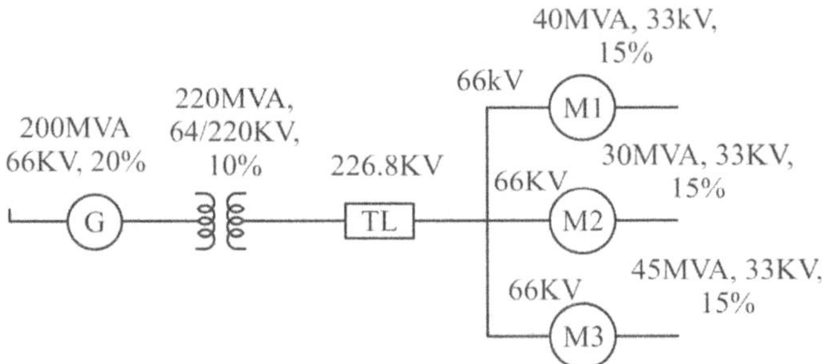

Fig.1.9

The base voltage in transmission line = 66 × (220/64)

$$= 226.875 \text{ KV}$$

In the motor circuit, The base voltage = 226.875 × (64/220)

$$= 66 \text{ KV}$$

Base impedance in the transmission line = (Base KV^2 / Base MVA) Ω

$$= ((226.875)^2/200) \text{ Ω}$$

$$= 257.36 \text{ Ω}$$

Per unit impedance = (j45 / j257.686) p.u.

$$= 0.1746 \text{ p.u}$$

Motor 1:

Rating: 40 MVA, 33 KV, $X'' = 0.15$

$$Z_{p.u.new} = Z_{p.u\ given} \times \left(\frac{\text{Base KV given}}{\text{Base KV new}}\right)^2 \times \left(\frac{\text{Base MVA new}}{\text{Base MVA given}}\right)$$

Zp.u new = 0.15 × [(33/ 66)^2 × (200/30)] p.u.

$$= 0.1875 \text{ p.u.}$$

Motor 2:

Rating: 30 MVA, 33KV, $X'' = 0.15$.

$Z_{\text{p.u. new}} = 0.15 \times (33/66)^2 \times (200/45)\text{p.u}$

$\qquad = 0.25 \text{ p.u}$

Motor 3:

Rating: 45 MVA, 33kV, $X'' = 0.15$

$Z_{\text{p.u (new)}} = 0.15 \times (33/66)^2 \times (200/45) \text{ p.u}$

$\qquad = 0.16 \text{ p.u}$

Transformer:

Rating: 220 MVA, 64kV, $X'' = 0.1$

$X''(T) = 0.1 \times (64/66)^2 \times (200/220) = 0.08548 \text{ p.u}$

Example 8: Draw the per unit impedance diagram for the power system below. Neglecting the reactance and use a base of 150 MVA, 110 KV in 40 ohms line. The rating of the generator, motor and transformers are:

Generator: 45 MVA, 32 KV, X = 15%

Motor: 42 MVA, 11 KV, X = 25%

Y-Y Transformer: 35 MVA, 22/110 KV, 13%

Y- Transformer: 30 MVA, 20/110 KV, 13%

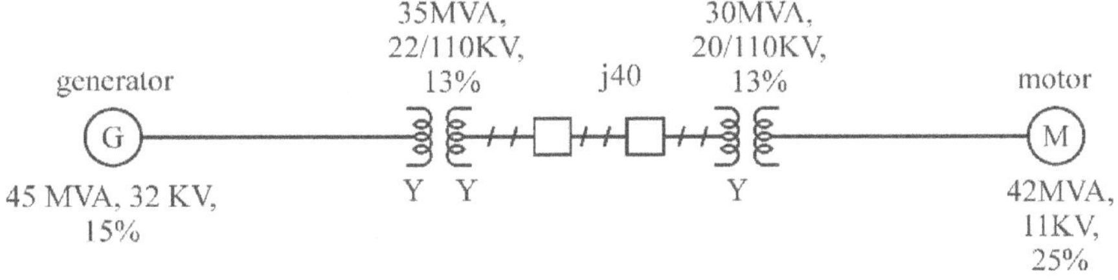

Fig.1.10

Solution:

$$Z_{\text{p.u(new)}} = Z_{\text{p.u(given)}} \times \left(\frac{\text{Base KV given}}{\text{Base KV new}} \right)^2 \times \left(\frac{\text{Base MVA new}}{\text{Base MVA given}} \right)$$

Generator:

New base voltage in generator = $(110 \times 22/110)$ KV = 22 KV

$X'' = 0.15 \times (32/22)^2 \times (150/45)\text{p.u} = 1.0578 \text{ p.u}$

Transformer 1:

New voltage in Transformer 1 = $(110 \times 22/110)$ KV = 22 KV

$X'' = 0.13 \times (22/22)^2 \times (150/35) \text{ p.u.} = 0.5571 \text{ p.u.}$

Transmission Line:

$$Z_B = \frac{\left(\text{Base KV}\right)^2}{\text{Base MVA}}$$

$$= (110)^2 / 150 \ \Omega$$

$Z_B = 80.66 \ \Omega$

$Z_{p.u.}$ = (Actual impedance / Base impedance) p.u.

$\qquad = 40 / 80.66$ p.u.

$\qquad = 0.49$ p.u.

Transformer 2:

New base voltage in Transformer 2 = $(110 \times 20 / 110)$ KV = 20 KV

$X'' = 0.13 \times (20 / 20)^2 \times (150 / 30)$ p.u.

$\qquad = 0.65$ p.u.

Motor:

$X'' = 0.25 \times (11 / 11)^2 \times (150 / 42)$ p.u.

$\qquad = 0.8928$ p.u.

Impedance diagram:

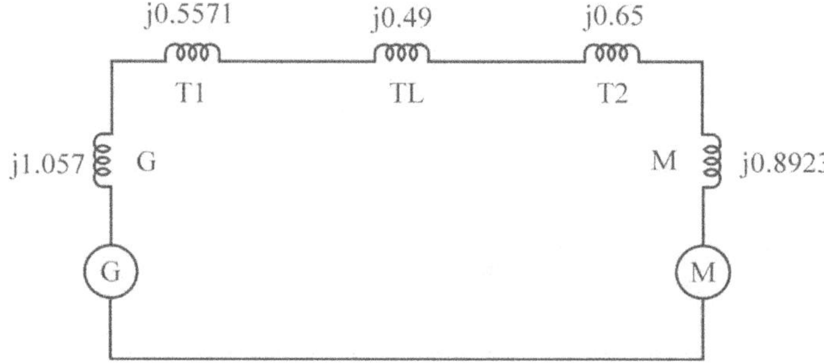

Fig.1.11

Example 9: 3 bus system is given in fig.1.12. The ratings of the various components are listed below:

Generator 1: 45 MVA, 12.5 KV, $X'' = 0.12$ p.u.

Generator 2: 30 MVA, 12.3 KV, $X'' = 0.3$ p.u.

Generator 3: 20 MVA, 10 KV, $X'' = 0.2$ p.u.

Transformer 1: 40 MVA, 11/110 KV Y/Y, X = 0.15 p.u.

Transformer 2: 20 MVA, 12.5/115 KV Y/Y, X = 0.2 p.u.

Transformer 3: 30 MVA, 12.5/115 KV Y/Y, X = 0.1 p.u.

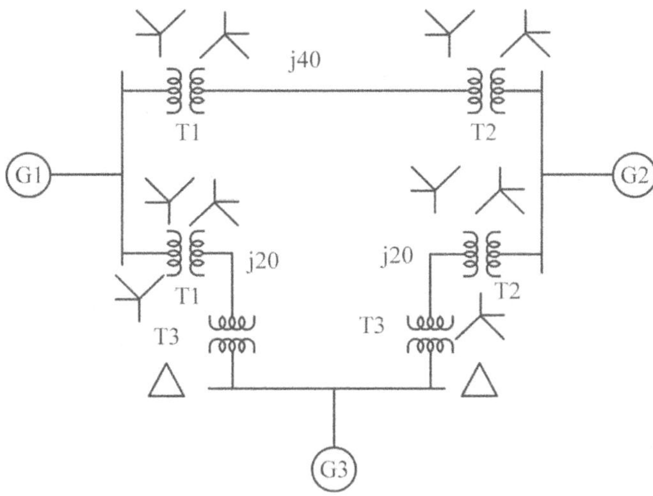

Fig.1.12

The line impedances are shown in Fig.1.12. Determine the reactance based on 45 MVA and 12.5 kV as base quantities in generator 1.

Solution:

Base MVA = 45 MVA; Base KV = 12.5 KV;

For Generator 1:

$$Z_{\text{p.u new}} = Z_{\text{p.u given}} \times \left(\frac{\text{Base KVgiven}}{\text{Base KVnew}}\right)^2 \times \left(\frac{\text{BaseMVAnew}}{\text{BaseMVAgiven}}\right)$$

$$Z_{\text{p.u new}} = 0.12 \times \left(\frac{12.5}{12.5}\right)^2 \times \left(\frac{45}{45}\right) = 0.12 \text{ p.u.}$$

Transformer 1:

40 MVA, 11 KV, X = 0.15p.u.

Base value 45 MVA, 12.5 KV

$$X_{T1} = Z_{\text{p.u new}} = 0.15 \times \left(\frac{11}{12.5}\right)^2 \times \left(\frac{45}{40}\right)$$

$$= 0.13068 \text{ p.u.}$$

Transformer 2:

20 MVA, 12.5 KV, X = 0.2 p.u.

Base voltage on the LV side of Transformer 2 = $12.5 \times \left(\frac{110}{11}\right)$

$$= 125 \text{ KV}$$

$$= 125 \times \left(\frac{12.5}{115}\right)$$

$$= 13.5 \text{ KV}$$

$$X_{T2} = 0.2 \times \left(\frac{12.5}{13.5}\right)^2 \times \left(\frac{45}{20}\right)$$

$$= 0.3858 \text{ p.u.}$$

Transformer 3:

Base voltage on the HV side of transformer 3 $= 12.5 \times \left(\frac{115}{12.5}\right)$

$$= 115 \text{KV}$$

$$= 15 \times \left(\frac{115}{12.5}\right)$$

$$= 138 \text{ KV}$$

$X = 0.1$ p.u, 115 KV, 30 MVA

$$X_{T3} = 0.1 \times \left(\frac{115}{138}\right)^2 \times \left(\frac{45}{30}\right) = 0.10416 \text{p.u}$$

Generator 2:

Base voltage $= 125 \times \left(\frac{12.5}{115}\right)$

$$= 13.5 \text{ KV}$$

$x'' = 0.3$ p.u., 12.3 KV, 30 MVA

$$X_{G2} = 0.3 \times \left(\frac{12.3}{13.5}\right)^2 \times \left(\frac{45}{30}\right)$$

$$= 0.3735 \text{p.u}$$

Generator 3:

Base voltage on the bus $= 138 \times \left(\frac{12.5}{115}\right) = 15 \text{ KV}$

20 MVA, 10 kV, $X'' = 0.2$ p.u.

$$X_{g3} = 0.2 \times \left(\frac{10}{15}\right)^2 \times \left(\frac{45}{20}\right)$$

$$= 0.2 \text{p.u.}$$

Transmission Line 1:

Base voltage along the transmission line whose impedance is $j40\Omega = 12.5 \times \left(\frac{110}{11}\right)$

$$= 125 \text{ KV}$$

$$\text{Per unit impedance} = \left(\frac{\text{Actual impedance}}{\text{Base impedance}}\right)$$

$$\text{Base impedance} = \frac{(\text{BaseKV})^2}{\text{BaseMVA}}$$

$$= \frac{(125)^2}{45}$$

$$= 347.22 \ \Omega$$

$$\text{Per unit impedance} = \left(\frac{40}{347.22}\right) = 0.1152 \text{ p.u.}$$

Transmission Line 2:

Base KV = 125, base MVA = 45

$$\text{Base impedance} = \frac{(125)^2}{45} = 347.22 \ \Omega$$

$$\text{Per unit impedance} = \left(\frac{20}{347.22}\right) = 0.0576 \text{ p.u.}$$

Transmission Line 3:

Base KV = 125, Base MVA = 45

$$\text{Base impedance} = \frac{(125)^2}{45} = 347.22 \ \Omega$$

$$\text{Per unit impedance} = \left(\frac{20}{347.22}\right) = 0.0576 \text{ p.u.}$$

Reactance Diagram:

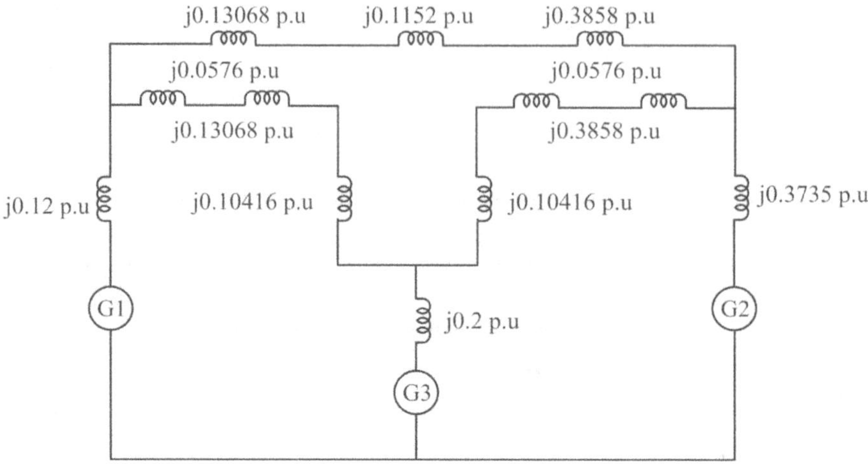

Fig.1.13

Example 10: 13MVA, 8 KV, 3-φ Generator has a subtransient reactance of 40%. It is connected through a Δ-Y transformer to a high voltage transmission line having a total series reactance of 60Ω.At the load end of the line is a Y-Y step down transformer .Both transformer bank are composed of 1- φ transformer connected for 3- φ operation. Each of the two transformer comprising each bank is rated 5050 KVA,11/110 KV with a reactance 10%. The load represented as a motor which is drawing 12MVA at 10.2KV with 60% reactance. Draw the impedance diagram showing all impedance in per unit. Choose a base of 12MVA, 12.5KV at the motor.

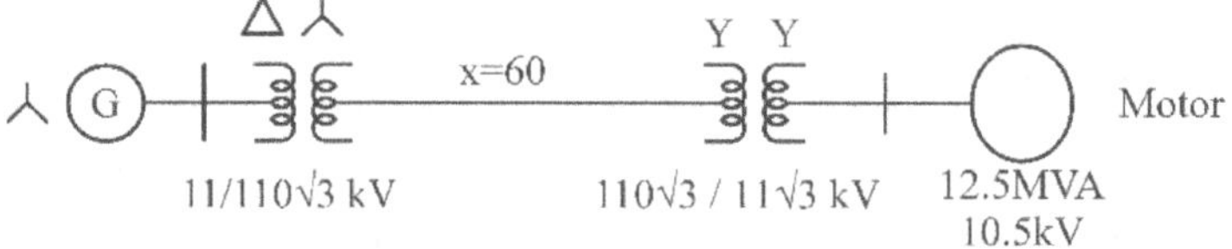

Fig.1.14

Solution:

Base MVA= 12; Base KV = 12.5

Load reactance = 0.6pu.

Total MVA rating of the transformer 1 and 2 separately = 3 x5050/1000

$$= 15.15MVA$$

Base KV on the transmission line side $=10.2 \times \left(\dfrac{110\sqrt{3}}{11\sqrt{3}} \right) = 102$ kV

$$\text{Base impedance} = \frac{(\text{BaseKV})^2}{\text{BaseMVA}}$$

Base impedance in the transmission line $=102^2/12= 867\Omega$

$$\text{Per unit impedance} = \left(\frac{\text{Actual impedance}}{\text{Base impedance}}\right)$$

Per unit impedance$=60/867=0.0653$ p.u

Transformer 1:

Base KV on the generator side is given by $=\left(\dfrac{11}{110\sqrt{3}}\right) \times 102 = 5.9$ KV

Per unit of transformer $X_{t_1} = 0.1 \times \left(\dfrac{11}{5.9}\right)^2 \times \left(\dfrac{12}{15.15}\right) = 0.2753$ p.u

Generator 1:

Given: 13 MVA, 8 KV, $X'' =0.4$ p.u.

Base value = 12 MVA, 5.9 KV

$$X_{g_1} = 0.4 \times \left(\frac{8}{5.9}\right)^2 \times \left(\frac{12}{13}\right) = j0.6788 \text{p.u}$$

Transformer 2:

Given:15.15 MVA, $110\sqrt{3}$ KV, $X'' =0.1$

Base value $=10.5$ MVA,102 KV,

$$X_{t_2} = 0.1 \times \left(\frac{110\sqrt{3}}{102}\right)^2 \times \left(\frac{10.5}{15.15}\right)$$

$$= 0.235 \text{ p.u.}$$

Impedance diagram:

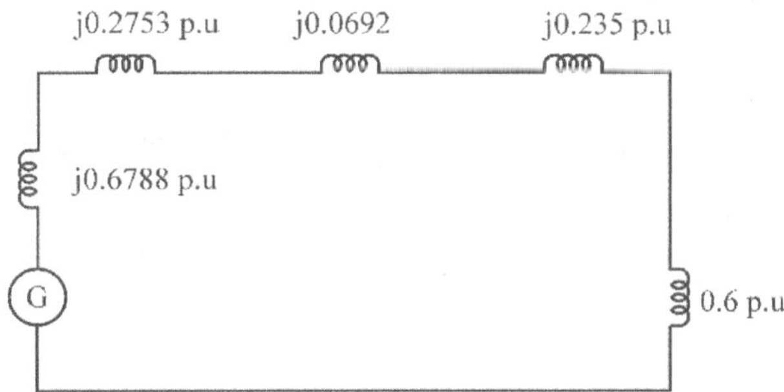

Fig.1.15

Example 11: Figure 1.16 shows a single line diagram of an unloaded three generator power system with interconnection between the generators by means of three transformers and a transmission line with two sections, with their impedances marked on the diagram. The ratings of the generators and transformers are given below:

Generator	MVA	KV	Reactance p.u
1	50	13.8	0.15
2	40	13.2	0.2
3	30	11	0.25

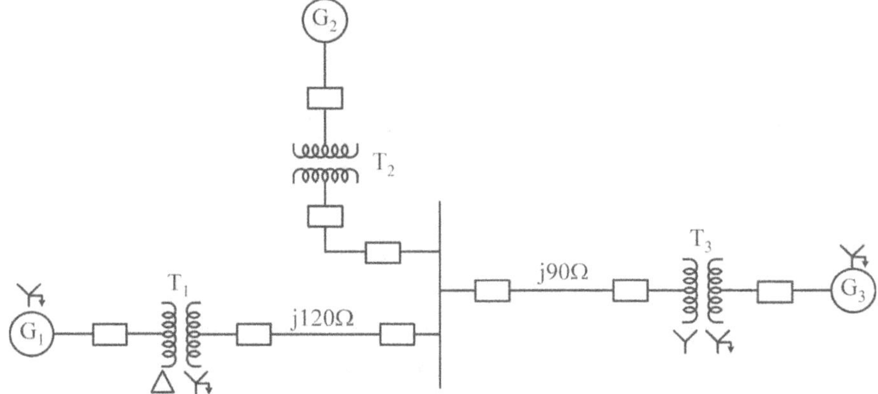

Fig.1.16

Transformer 1: 45MVA, 11Δ–110Y KV, X = 10%

Transformer 2: 25MVA, 12.5(del) – 115Y KV, X = 15%

Transformer 3: Single phase units, each rated 10 MVA, 6.9/69 KV, X = 10%

Draw the impedance diagram and mark all values in p.u. choosing a base of 30MVA, 13.8KV in the generator 1 circuit.

Solution:

Base MVA = 30; Base KV = 13.8KV

Generator 1:

$$Z_{p.u.new} = Z_{p.u.given} \times \left(\frac{\left(\text{Base KV}_{given} \right)}{\text{Base KV}_{new}} \right)^2 \times \left(\frac{\left(\text{Base MVA}_{new} \right)}{\text{Base MVA}_{given}} \right)$$

$$X_{g1} = 0.15 \times (13.8/13.8)^2 \times 30/50$$

$$X_{g1} = 0.09 \text{ p.u.}$$

Transformer 1:

Given : 45MVA, 11Δ /110Y KV, X = 10%

$X_{T1} = 0.1 \times (11/13.8)^2 \times (30/45)$

$X_{T1} = 0.042$ p.u.

Transmission Line 1:

Base voltage on the transmission line with impedance

$j120\Omega = 13.8 \times (110/11) = 138$ KV

$$\text{Base Impedance} = \frac{\left(\text{Base KV}\right)^2}{\text{Base MVA}}$$

$$\text{Base impedance} = \frac{\left(138\right)^2}{30} = 6.348\Omega$$

$$\text{Per unit impedance} = \frac{\text{Actual Im pedance}}{\text{Base Im pedance}}$$

Per unit impedance = j120/634.8 = j0.18903 p.u.

Transmission Line 2:

Per unit impediance = j90/634.8 = 0.14177 p.u.

Transformer 2:

Given: 25 MVA, 12.5 Δ /115Y KV, X = 0.15 p.u.

Base value: 30MVA, 13.8KV

$X_{T2} = 0.15 \times (115/13.8)^2 \times 30/25$

$X_{T2} = 12.5$ p.u.

Generator 2:

Base KV = 13.8 $\times$ 12.5/115 = 1.5 p.u.

Given : 25 MVA, 1.5 KV, X = 0.15 p.u.

$X_{G2} = 0.15 \times (1.5/1.5)^2 \times 30 / 25$

$\qquad = 0.18.$ p.u.

Transformer 3:

Line to line voltage when three single phase transformers are used as a three phase transformer is

$\sqrt{3} / \sqrt{3} \times 6.9/6.9 = 11.95$KV/1119.51KV

Given: 30MVA, 11.95/119.51 KV, X = 0.1p.u

Base values: 30MVA, 13.8 KV

$X_{T3} = 0.1 \times (119.51/13.8)^2 \times 30/30$

$X_{T3} = 0.499$ p.u.

Generator 3:

Base KV = $13.8 \times 11.95 / 119.51 = 13.79$ KV

Given: 30MVA, 11KV, X = 0.25

Base value: 30, 1.379

$X = 0.25 \times (11 /1.379)^2 \times 30/30$

$X = 1012.1$ p.u.

Impedance Diagram:

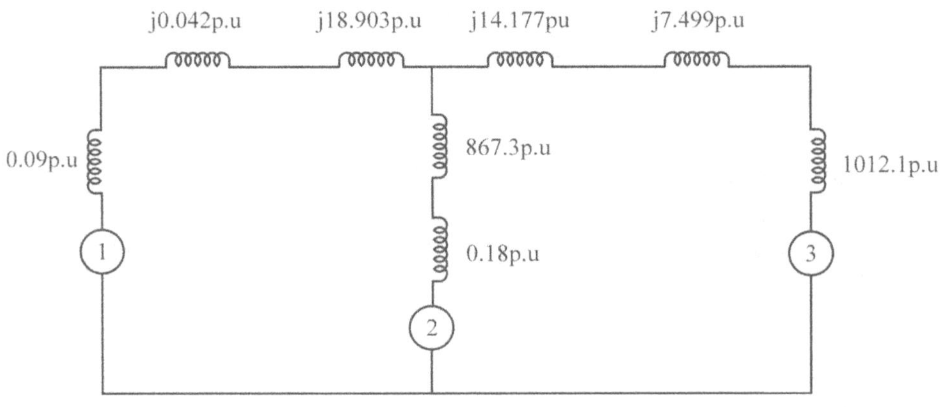

Fig.1.17

Example 12: The one line diagram of an unloaded power system is shown in Fig. 1.18. Reactances of the sections of transmission line are shown in the diagram. The generators and transformers are rated as follow:

Generator 1: 20 MVA, 13.8 KV, $X'' = 0.2$p.u

Generator 2: 30 MVA, 18 KV, $X'' = 0.2$p.u.

Generator 3: 30 MVA, 20 KV, $X'' = 0.2$p.u.

Transformer T1: 20 MVA, 20 KV, $X'' = 0.2$p.u.

Transformer T2: Single phase units each rated 30MVA, 127/18KV, X = 10%

Transformer T3: 35MVA, 220Y/22Y KV, X = 10%

Draw the impedance diagram with all reactance's marked in p.u. and with letters to indicate points corresponding to the one–line diagram. Choose a base of 50MVA, 13.8 KV in the circuit of generator 1.

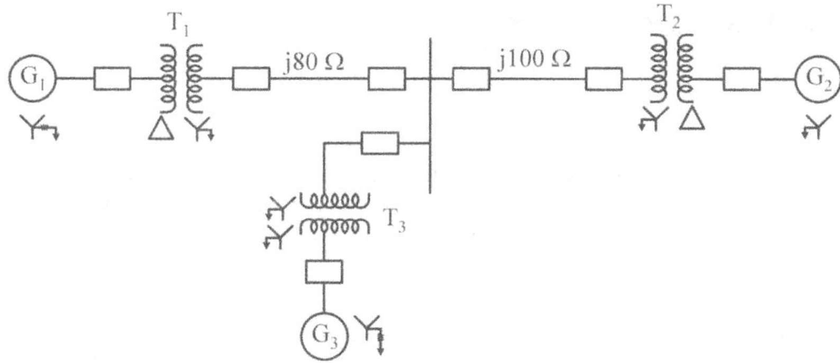

Fig.1.18

Solution:

50MVA, 13.8KV – Base values

Generator 1:

20MVA, 13.8KV, $X'' = 0.2$.

$$Z_{p.u.new} = Z_{p.u.given} \times \left(\frac{\left(Base\ KV_{given} \right)}{Base\ KV_{new}} \right)^2 \times \frac{Base\ MVA_{new}}{Base\ MVA_{given}}$$

$X_{G1} = 0.2 \times (13.8/13.8)^2 \times 50\,/20$

$\qquad = 0.5$ p.u.

Transmission line 1:

Base voltage on the transmission line using the transformation ratio = 220kV

$$Base\ impedance = \frac{\left(Base\ KV \right)^2}{Base\ MVA}$$

Base impedance = $(220 \times 220)/50 = 968\Omega$

$$Per\ unit\ impedance = \frac{Actual\ Impedance}{Base\ Impedance}$$

Per unit impedance = 80/968 = 0.0826 p.u.

Transmission line 2:

Per unit impedance = 100 / 968 = 0.1033 p.u.

Transformer 1:

Base MVA = 50, KV = 13.8

Given MVA = 25.13 MVA, 13.8KV, X = 10p.u.

$$X_{t1} = 0.1 \times (13.8 \ / \ 13.8)^2 \times 50 \ /25$$
$$= 0.2 \ \text{p.u.}$$

Transformer 2:

Voltage rating of the transformers when they are put to form a three phase transformer

$$= \sqrt{3} \times 127/18 \ \text{KV} = 220/18 \ \text{KV}$$

$$X_{t2} = 0.1 \times (18/18)^2 \times 50/30$$
$$= 0.166 \ \text{p.u.}$$

Generator 2:

Given: 30MVA, 18KV, $X'' = 0.2$

Base values: 50 MVA, 18KV

$$X_{g2} = 0.2 \times (18/18)^2 \times 50/30$$
$$= 0.3333 \ \text{p.u.}$$

Generator 3:

Given: 30MVA, 20KV, $X'' = 0.2$ p.u.

Base Values: 50MVA, 22KV

$$X_{g3} = 0.2 \times ((20/22)^2 \times 50/30$$
$$= 0.275 \ \text{p.u.}$$

Transformer 3:

$$X_{t3} = 0.1 \times 50/35 = 0.1429 \ \text{p.u.}$$

Impedance diagram:

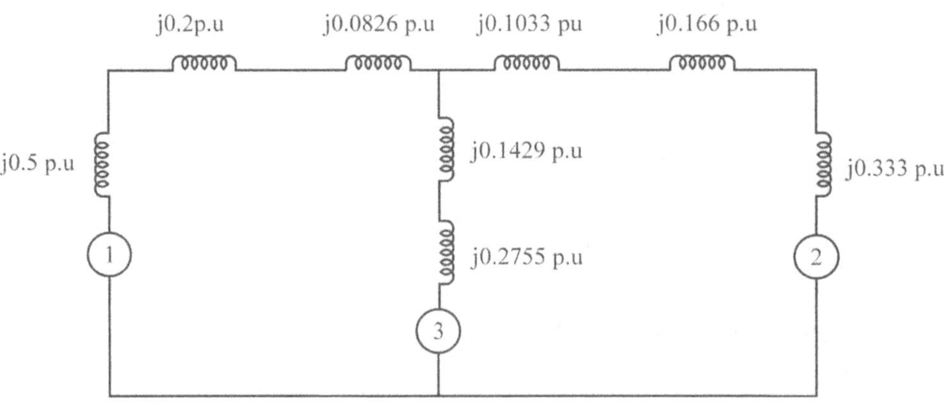

Fig.1.19

Example 13: Figure 1.20 shows a single line diagram of an unloaded three generator power system with interconnection between the generators by means of three transformers and a transmission line with two sections, with their impedances marked on the diagram. The rating of the generator and transformer are given below:

Generator	MVA	KV	Reactance p.u.
1	30	7.7	0.3
2	20	7.7	0.20
3	35	15.4	0.20

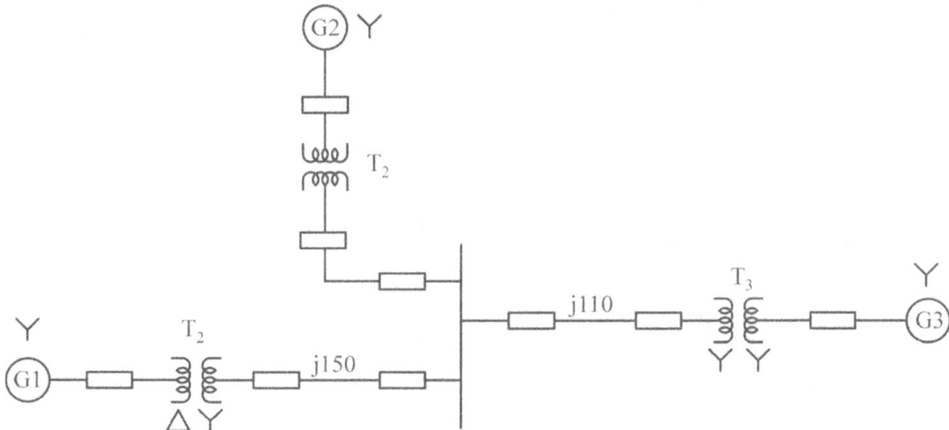

Fig.1.20

Transformer 1: 35 MVA, $8.0\Delta - 120Y$, X = 10%

Transformer 2: 20 MVA, $8.0\Delta - 120Y$, X = 10%

Transformer 3: Single phase units, each rated 15 MVA, 8.0/80KV, X = 10%

Draw an impedance diagram and mark all values in p.u. choosing a base of 35MVA, 7.7KV in generator 1 circuit.

Solution:

Base MVA=35; Base KV=7.7KV

Generator 1:

$$Z_{p.u\ new} = Z_{p.u\ given} \times \left(\frac{\text{Base KV}_{given}}{\text{Base KV}_{new}}\right)^2 \times \left(\frac{\text{BaseMVA}_{new}}{\text{BaseMVA}_{given}}\right)$$

$$X_{G1} = 0.3 \times \left(\frac{7.7}{7.7}\right)^2 \times \frac{35}{30} = 0.35\text{p.u.}$$

Transformer 1:

Given: 35MVA, 8Δ – 120Y KV, X = 10%

$$X_{T1} = 0.1 \times \left(\frac{8.0}{7.7}\right)^2 \times \frac{35}{35} = 0.1 \times 1.0794 = 0.10794 \text{p.u.}$$

Transmission line 1:

Base voltage on the transmission line with impedance j150 Ω = $7.7 \times \dfrac{12}{8}$ = 115.5

$$\text{Base Impedance} = \left(\frac{\text{BaseKV}}{\text{BaseMVA}}\right)^2$$

$$\text{Base Impedance} = \frac{(115.5)^2}{35} = 381.15 \ \Omega$$

$$\text{Per unit impedance} = \left(\frac{\text{Actual impedance}}{\text{Base impedance}}\right)$$

$$\text{Per unit impedance} = \frac{\text{j}150}{381.15} = \text{j}0.39354 \text{p.u.}$$

Transmission line 2:

$$\text{Per unit impedance} = \frac{\text{j}110}{381.15} = \text{j}0.2886 \text{p.u.}$$

Transformer 2:

Given: 20MVA, 8.0Δ – 120Y KV, X = 0.1p.u.

Base values: 35 MVA, 115.5 KV

$$X_{T2} = 0.1 \times \left(\frac{120}{115.5}\right)^2 \times \frac{35}{20}$$

$$= 0.1889 \text{p.u.}$$

Generator 2:

$$\text{Base KV} = 115.5 \times \frac{8}{120} = 7.7 \text{ KV}$$

Given: 20 MVA, 7.7 KV, X = 0.20p.u.

$$X_{G2} = 0.2 \times \left(\frac{7.7}{7.7}\right)^2 \times \frac{35}{20} = 0.35 \text{p.u.}$$

Transformer 3:

line to line voltage when three 1-Ø transformer are used as a 3-Ø transformer is

$$\frac{\sqrt{3}}{\sqrt{3}} \times \frac{8}{80} = \frac{13.856\text{kV}}{138.564\text{kV}}$$

Given: 45MVA, 13.865/138.564 KV, X = 0.1p.u.

Base values: 35MVA, 115.5KV

$$X_{T3} = 0.1 \times \left(\frac{(138.56)}{115.5}\right)^2 \times \frac{35}{45} = 0.1119 \text{ p.u.}$$

Generator 3:

$$\text{Base KV} = 115.5 \times \frac{13.856}{138.564} = 11.549\text{KV}$$

Given: 35MVA, 15.4KV, X = 0.2p.u.

Base values: 35MVA, 11.549KV

$$X_{G3} = 0.2 \times \left(\frac{15.4}{11.549}\right)^2 \times \frac{35}{35} = 0.3556\text{p.u.}$$

Impedance diagram:

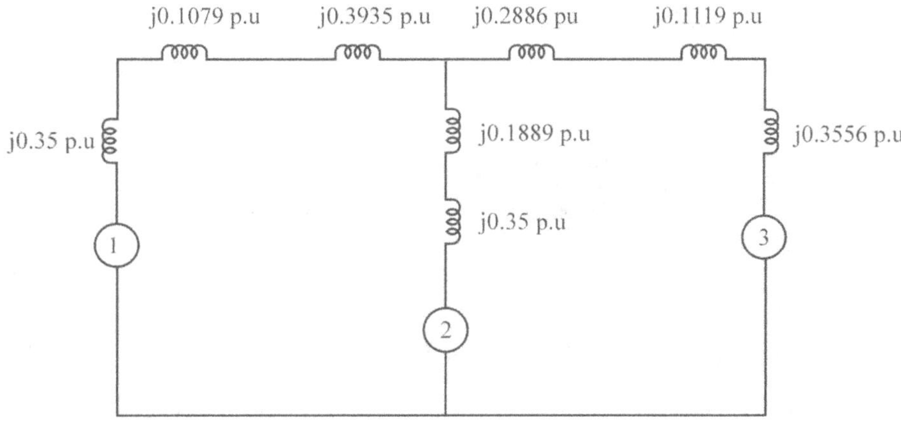

Fig.1.21

Example 14: The one line diagram of an unloaded power system is shown in Fig. 1.22. Reactance of two sections of transmission line are shown in the diagram. The generators and transformers are rated as follows:

Generator 1: 30MVA, 15.8KV, X″ = 0.2p.u.

Generator 2: 40MVA, 20KV, X″ = 0.2p.u.

Generator 3: 40MVA, 22KV, $X'' = 0.2$ p.u.

Transformer 1: 30MVA, 220KV, $X'' = 10\%$

Transformer 2: Single phase unit each rates 10MVA, 130/20 KV, $X = 10\%$

Transformer 3: 40MVA, 220Y/25Y KV, $X = 10\%$

Draw the impedance diagram with all reactance marked in p.u. and with letters to indicate points corresponding to the one-line diagram. Choose a base of 60MVA, 15.8KV in the circuit of generator 1.

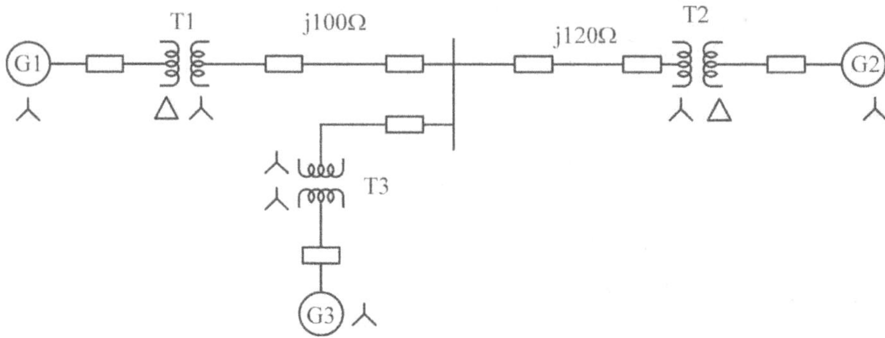

Fig.1.22

Solution:

Base values: 60MVA, 15.8KV

Generator 1:

30MVA, 15.8KV, $x'' = 0.2$

$$Z_{\text{p.u new}} = Z_{\text{p.u given}} \times \left(\frac{\text{Base KV}_{\text{given}}}{\text{Base KV}_{\text{new}}} \right) \times \left(\frac{\text{BaseMVA}_{\text{new}}}{\text{BaseMVA}_{\text{given}}} \right)$$

$$X_{G1} = 0.2 \times \left(\frac{15.8}{15.8} \right)^2 \times \left(\frac{60}{30} \right)$$

$$= 0.4 \text{p.u.}$$

Transmission Line 1:

Base voltage on the transmission line using the transformation ratio = 220 KV

$$\text{Base impedance} = \frac{(\text{Base KV})^2}{\text{Base MVA}}$$

$$\text{Base impedance} = \frac{(220)^2}{60}$$

$$= 806.67 \Omega$$

$$\text{Per unit impedance} = \left(\frac{\text{Actual impedance}}{\text{Base impedance}} \right)$$

$$= \frac{100}{806.67} = 0.1239 \text{p.u.}$$

Transmission Line 2:

$$\text{Per unit impedance} = \frac{120}{806.67} = 0.1487 \text{p.u.}$$

Transformer 1:

Base values: 60MVA, 15.8KV

Given: 30MVA, 15.8KV, X = 10%

$$Z_{\text{p.u new}} = Z_{\text{p.u given}} \times \left(\frac{(\text{BasekVgiven})}{\text{BasekVnew}} \right)^2 \times \left(\frac{\text{BaseMVAnew}}{\text{BaseMVAgiven}} \right)$$

$$X_{T1} = 0.1 \times \left(\frac{15.8}{15.8} \right)^2 \times \left(\frac{60}{30} \right)$$

$$X_{T1} = 0.2 \text{p.u}$$

Transformer 3:

$$X_{T3} = 0.1 \times \left(\frac{220}{220} \right)^2 \times \left(\frac{60}{40} \right)$$

$$= 0.15 \text{p.u}$$

Transformer 2:

Voltage of the transformer when they are put to form a 3-Ø transformer

$$= \frac{\sqrt{3}}{220} \times 130 \text{ KV} = \frac{225}{220} \text{KV}$$

$$X_{T2} = 0.1 \times \left(\frac{225}{220} \right)^2 \times \left(\frac{60}{30} \right) = 0.209 \text{p.u.}$$

Generator 2:

Given: 40MVA, 20KV, $X'' = 0.2$

Base values: 60MVA, 19.5KV

$$X_{G2} = 0.2 \times \left(\frac{20}{19.5} \right)^2 \times \left(\frac{60}{40} \right) = 0.315 \text{p.u.}$$

Generator 3:

Given: 40MVA, 22KV, $X'' = 0.2$

Base values: 60MVA, 25KV

$$X_{g3} = 0.2 \times \left(\frac{22}{25}\right)^2 \times \left(\frac{60}{40}\right) = 0.232 \text{p.u.}$$

Impedance diagram:

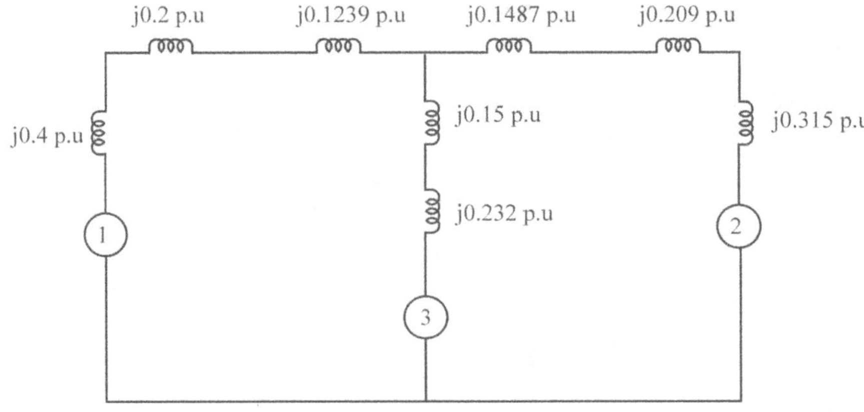

Fig.1.23

Example 15: A 3 bus system is given in Fig.1.24. The ratings of the various components are listed below:

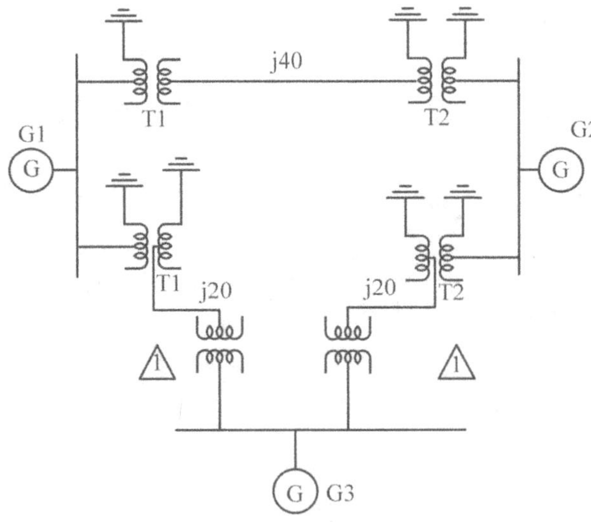

Fig.1.24

Generator 1: 40 MVA, 13.8 KV, $X'' = 0.2$p.u.

Generator 2: 50 MVA, 14.2 KV, $X'' = 0.15$p.u.

Generator 3: 30 MVA, 12 KV, $X'' = 0.25$p.u.

Transformer 1: 50 MVA, 10 KV/100KV Y/Y, $X = 0.12$p.u.

Transformer 2:30 MVA, 11.5 KV/110KV Y/Y, $X = 0.14$p.u.

Transformer 3:35 MVA, 11.5 KV/110KV Y/Y, $X = 0.2$p.u.

The line impedances are shown in the figure. Draw the reactance diagram based on 40 MVA and 13.8 kV as base quantities in Generator 1.

Solution:

Base MVA = 40; Base KV = 13.8

Generator 1:

$$Z_{\text{p.u.new}} = Z_{\text{p.u.given}} \times \left(\frac{\text{Base KV}_{\text{given}}}{\text{Base KV}_{\text{new}}} \right)^2 \times \frac{\text{Base MVA}_{\text{new}}}{\text{Base MVA}_{\text{given}}}$$

Per unit reactance = $0.2 \times (13.8/13.8)^2 \times 40/40 = 0.2$p.u.

Transmission line 1:

Base voltage along the transmission line whose impedance is $j40\Omega = 13.8 \times 100/10 = 138$KV

$$\text{Base impedance} = \frac{(\text{Base KV})^2}{\text{Base MVA}}$$

$$\text{Base impedance} = \frac{138^2}{40} = 476.1\Omega$$

$$\text{Per unit impedance} = \frac{\text{Actual impedance}}{\text{Base impedance}}$$

Per unit impedance = $40/476.1 = 0.0840$p.u.

Generator 2:

Base Voltage = $138 \times 11.5/110 = 14$KV

$$X_{G2} = 0.15 \times (14.2/14)^2 \times 40/50$$

$$= 0.15 \times (14.2/14)^2 \times 0.8 = 0.1234\text{p.u.}$$

Transformer 1:

50 MVA, 10 KV $X = 0.12$p.u.

Base values: 40 MVA, 13.8 KV

$X_{T1} = 0.12 \times (10/13.80)^2 \times 40/50 = 0.05040$p.u.

Transformer 2:

30 MVA, 11.5 KV X = 0.14p.u.

Base voltage on the L.T. side of the transformer 2 = 138 × 11.5/110 = 14KV

X_{T2} = 0.14 × (11.5/14)² × 40/30 = 0.14 × 0.674744 × 1.3333 = 0.1259p.u.

Transformer 3:

Base voltage on H.T. side of the transformer 3 is 138 KV.

X_{T3} = 0.2 × (110/138)² × 40/35

= 0.2 × 1.428571429 × 12100/19044

= 0.14522p.u.

Generator 3:

Base voltage on the bus= 138 × (11.5/110) =14KV

30 MVA, 12KV, X″ = 0.25p.u.

Base values: 40 MVA, 14KV

= 0.25 × (12/14)² × 40/30

= 0.25 × 144/196 × 1.3333

= 0.244p.u.

Transmission Line 2:

Base KV = 138

Base MVA = 40

Base impedance = (138)²/40 = 476.1Ω

Per unit impedance = j20/476.1 = j0.04200

Reactance Diagram

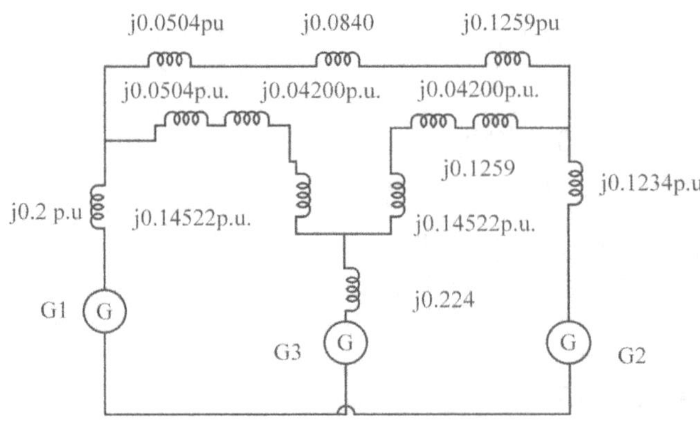

Fig.1.25

Example 16: A 20 MVA, 9.5KV 3-ϕ Generator has a sub-transient reactance of 20%. It is connected through a Δ-Y transformer to a high voltage transmission line having a total series reactance of 60Ω . At the load end of the line is a Y-Y step down transformer .Both transformers banks are composed of 1-ϕ transformers connected for 3-ϕ operation. Each of the two transformers comprising each bank is rated 5667 KVA, 11/110 KV with a reactance of 10 %. The load represented as impedance is drawing 11 MVA, 13.5 KV at 70 % power factor lagging. Draw the impedance diagram showing all impedances in per unit. Choose a base   of 11 MVA, 13.5KV in the load circuit.

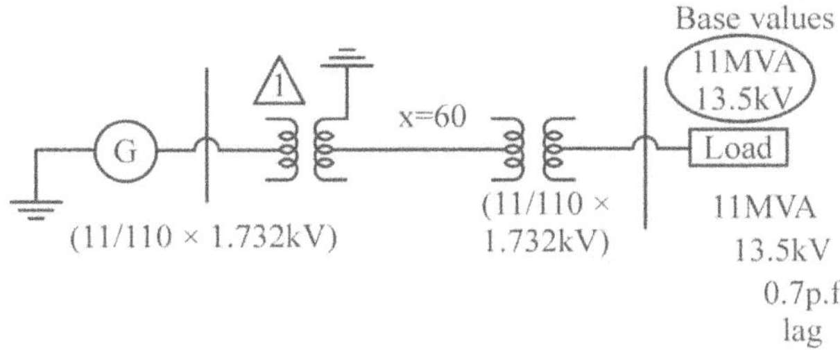

Fig.1.26

Solution:

MVA = 11; KV = 13.5

Load impedance = 0.7 + j0.6p.u. (Using the formula for p.u. impedance)

Total MVA rating of the transformers 1 and 2 separately = 3 × 5667/1100 = 15.4545 MVA

Base KV on the transmission line side = 13.5 × 110 $\sqrt{3}$ /11 $\sqrt{3}$ = 135 KV

Base impedance = (Base KV)^2/Base MVA

Base impedance in the transmission line side = (115)^2/11 = 1656.8181

Per unit impedance = $\dfrac{\text{Actual impedance}}{\text{Base impedance}}$

Per unit impedance = 60/1656.8181 = 0.036213

Transformer 1:

Base KV on the generator side is given by

$= 115 \times 11/110\sqrt{3}$ = 6.6 KV

Therefore, per unit impedance of transformer 1 with reference to low voltage winding is given by

$X_{T1} = 0.1 \times (11/6.6)^2 \times 11/15 = 0.2037$p.u.

Generator 1:

Given, 20 MVA, 9.5 KV, $X'' = 0.2$

Base values: 11 MVA, 6.6 KV

$$X'' = 0.2 \times (110\sqrt{3}/115)^2 \times 11/15 = 0.4025 \text{p.u.}$$

Transformer 2:

Given 15 MVA, $110\sqrt{3}$ KV, $X'' = 0.3$p.u.

Base values: 11 MVA, 115 KV

$$X_{T2} = 0.3 \times (110\sqrt{3}/115)^2 \times 11/15 = 0.3 \times 36300/13225 \times 0.73333 = 0.06038 \text{p.u.}$$

Impedance Diagram:

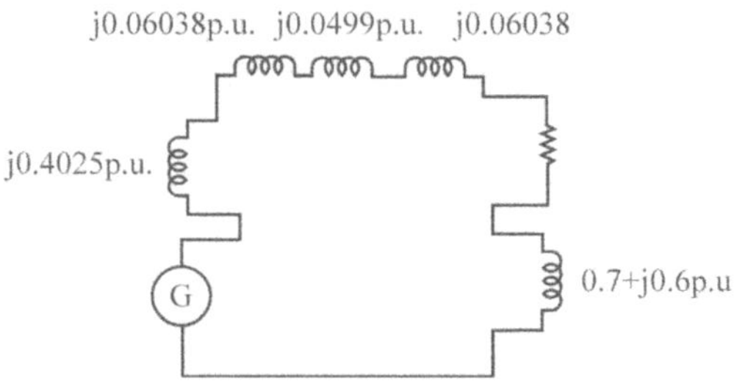

Fig.1.27

Example 17: Fig.1.28 shows a single line diagram of an unloaded three generator power system with interconnection between the generators by means of three transformers and a transmission line with two sections, with their impedances marked on the diagram. The rating of the generators and transformers are given below:

Generator 1: 30MVA, 6.9KV, $X'' = 20$p%.

Generator 2: 20MVA, 6.9KV, $X'' = 10$%.

Generator 3: 35MVA, 13.4KV, $X'' = 10$%.

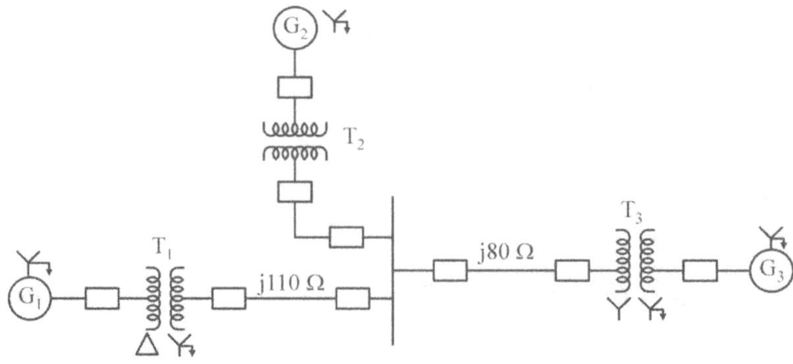

Fig.1.28

Transformer 1: 35MVA, 7.2 Δ-120YKV, X = 10%

Transformer 2: 20MVA, 7.2 Δ -120 YKV, X = 10%

Transformer 3: Single phase units, each rated 10MVA, 7.9/79KV, X = 10%

Draw an impedance diagram and mark all values in p.u. choosing a base of 35MVA, 6.9KV in the generator 1 circuit.

Solution:

Take Base MVA = 35 and Base KV = 6.9

Generator 1:

$$X_{G1} = 0.2 \times (35/30) \times (6.9 \times 6.9)^2 = 0.2 \times (35/30) = 0.233 \text{p.u.}$$

For Transformer 1:

$$X_{T1} = 0.1 \times (35/35) \times (7.2/6.9)^2 = 0.1088 \text{ p.u.}$$

Transmission Line 1:

Base Voltage with impedance j110 Ω = 6.9 × (120/7.2) = 115KV

Now, $$Z_{\text{p.u.}} = Z_{\text{actual}} \times \left\{ \frac{\text{Base MVA}}{\left(\text{Base KV}\right)^2} \right\}$$

$$= 110 \times \{35/(115)^2\} = j0.2911 \text{ p.u.}$$

Transmission Line 2:

Base Voltage with impedance j80 Ω = 115KV

Now, Per unit impedance $$= Z_{\text{actual}} \times \left\{ \frac{\text{Base MVA}}{\left(\text{Base KV}\right)^2} \right\}$$

$$= 80 \times \{35/(115)^2\} = j0.2117 \text{ p.u.}$$

Transformer 2:

$$X_{T2} = 0.1 \times (35/20) \times (120/115)^2 = 0.1905 \text{ p.u.}$$

Generator 2:

Base KV = (115 × 7.2)/120 = 6.9KV

$$X_{G2} = 0.1 \times (35/20) \times (6.9/6.9)^2 = 0.175 \text{ p.u.}$$

Transformer 3:

Line to Line Voltage when three $1 - \varnothing$ Transformers are used as a $3 - \varnothing$ transformer is

$$(79 \times \sqrt{3})/(7.9 \times \sqrt{3}) = 136.83/13.68 \text{KV}$$

Base MVA in $3 - \varnothing$ = 10 × 3 = 30MVA

$$X_{T3} = 0.1 \times (35/10) \times (136.83/115)^2 = 0.4954 \text{ p.u.}$$

Generator 3:

Base KV = (115 × 13.68)/136.83 = 11.49KV

$$X_{G3} = 0.1 \times (35/35) \times (13.4/11.49)^2 = 0.1360 \text{ p.u.}$$

Impedance Diagram:

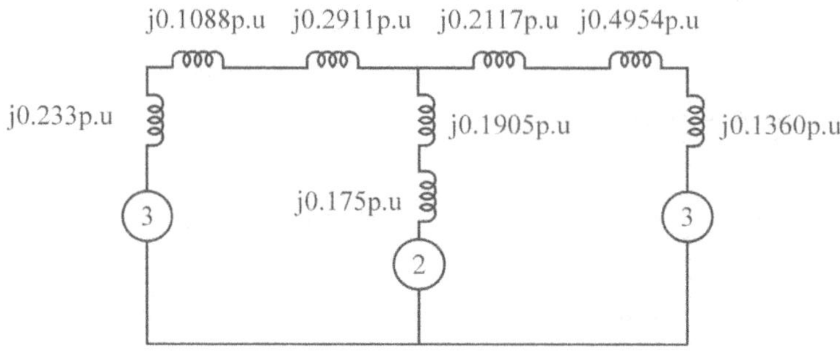

Fig.1.29

Example 18: The one line diagram of an unloaded power systems is shown in Fig.1.30. Reactance's of the two sections of transmission line are shown in the diagram. The generators and transformers are rated as follows:

G_1: 25MVA, 14.8KV, $X'' = 0.2$ p.u.

G_2: 35MVA, 19KV, $X'' = 0.2$ p.u.

G_3: 35MVA, 21KV, $X'' = 0.2$ p.u.

T_1 : 30MVA, 220KV, X = 10%

T_2 : Single Phase units each rated 10MVA, 127/18KV, X = 10%

T_3 : 40MVA, 220Y/21Y KV, X = 10%

Draw the impedance diagram with all reactance's marked in p.u. and with letters to indicate points corresponding to the one-line diagram. Choose a base of 60MVA, 14.8KV in the circuit of generator 1.

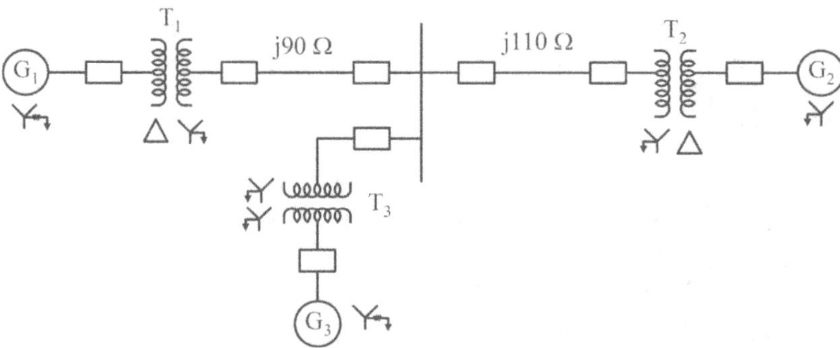

Fig.1.30

Solution:

50MVA, 14.8KV – Base Values

$$G_1:\ X_{G1} = 0.2 \times \left(\frac{50}{25}\right) \times \left(\frac{14.8}{14.8}\right)^2 = 0.4\ \text{p.u.}$$

$$T1: X_{T2} = 0.1 \times \left(\frac{50}{30}\right) \times \left(\frac{14.8}{14.8}\right)^2 = 0.16\ \text{p.u.}$$

TL$_1$:

$$\text{Base voltage on the transmission line} = 14.8 \times \left(\frac{220}{14.8}\right) = 220\text{KV}$$

$$\text{Per unit impedance} = 90 \times \left(\frac{50}{220 \times 220}\right) = 0.0929\ \text{p.u.}$$

TL$_2$:

$$\text{Per unit analysis} = 110 \times \left(\frac{50}{220 \times 220}\right) = 0.1136\ \text{p.u.}$$

T$_2$: Voltage rating of the transformers when they are put to form a 3–∅ transformer

$$= \frac{\sqrt{3} \times 127}{18}\text{kV} = \frac{207.84}{18}\text{kV}$$

MVA given for 3–∅ will be $10 \times 3 = 30$ MVA

$$X_{T2} = 0.1 \times \left(\frac{50}{30}\right) \times \left(\frac{127 \times \sqrt{3}}{220}\right)^2 = 0.16661\ \text{p.u.}$$

G$_2$: Base voltage $= (220 \times 18)/(127 \times \sqrt{3}) = 19.05$ KV

$$X_{G2} = 0.2 \times \left(\frac{50}{35}\right) \times \left(\frac{19}{19.05}\right)^2 = 0.2842\ \text{p.u.}$$

For T$_3$:

Base Voltage = 220KV

$$X_{T3} = 0.1 \times \left(\frac{50}{40}\right) \times \left(\frac{220}{220}\right)^2 = 0.125\ \text{p.u.}$$

G3: Base Voltage $= (220 \times 21)/220 = 21$KV

$$X_{G3} = 0.2 \times \left(\frac{50}{35}\right) \times \left(\frac{21}{21}\right)^2 = 0.2857 \text{ p.u}$$

Reactance Diagram:

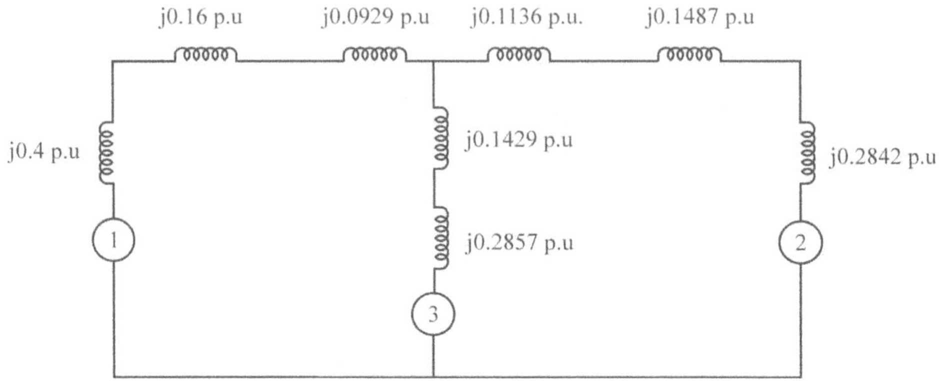

Fig.1.31

Example 19: Two Generators rated at 11.2 MVA, 13 KV and 15 MVA, 13.2 KV are connected in parallel to a bus bar. They feed supply to two motors of input 10 MVA, and 12 MVA respectively. The Operating Voltage of motors is 11.2KV. Assuming base quantities as 40 MVA and 14.8 KV, Draw the reactance diagram. The percentage Reactance for Generators is 13% and that for motors is 18%.

Solution:

Generator 1:

$Z_{p.u.given}$ = 0.13, Base KV$_{given}$ =13KV, Base KV Actual = 14.8 KV,

Base MVA$_{Actual}$ = 40MVA, Base MVA $_{Given}$ = 11.2MVA

$Z_{p.u. new}$ = 0.13 × (13/14.8)^2 × 40/11.2 = 0.358218 p.u.

Generator 2:

$Z_{p.u.given}$ = 0.13, Base KV $_{given}$ = 13.2 KV, Base KV $_{Actual}$ = 14.8 KV,

Base MVA$_{Actual}$ = 40MVA, Base MVA$_{given}$ = 15MVA

$Z_{p.u.new}$ = 0.13 × (13.2/14.8)^2 × 40/15 = 0.275763 pu

Motor 1:

$Z_{p.u.given}$ = 0.18, Base KV $_{given}$ = 11.2KV, Base KV$_{Actual}$ = 14.8KV,

Base MVA$_{Actual}$ = 40MVA, Base MVA$_{given}$ = 10MVA

$Z_{p.u.new}$ = 0.18 × (11.2/14.8)^2 × 40/10 = 0.41233 p.u.

Motor 2:

$Z_{p.u.given}$ = 0.18, Base KV$_{given}$ = 11.2KV, Base KV Actual = 14.8KV,

Base MVA $_{Actual}$ = 40 MVA, Base MVA$_{given}$ = 13.5 MVA

Zp.u. new = $0.18 \times (11.2/14.8)^2 \times 40/13.5 = 0.305429$ p.u.

Reactance Diagram:

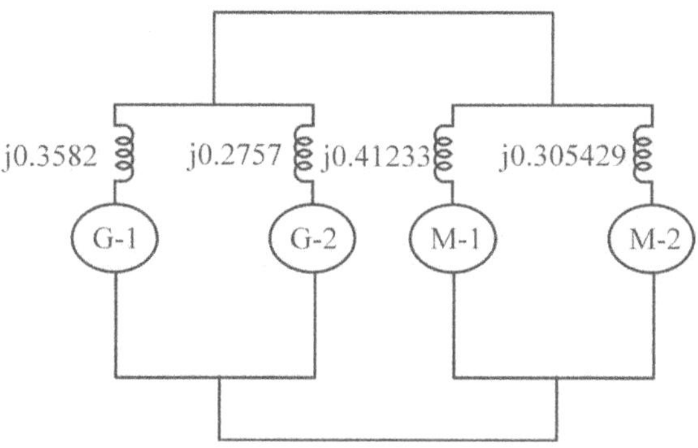

Fig.1.32

Example 20: Draw the Impedance Diagram for the power system shown in fig.1.33 impedance's are in pu. Neglect resistance and use a base of 40 KVA, 120 KV, 40 Ω line. The Ratings of the generators, motors and transformers are

Generator 1: 25 MVA, 15 KV, Reactance % = 20%

Generator 2: 28 MVA, 16 KV, Reactance % = 18%

Synchronous motor 3: 33 MVA, 14.8 KV, Reactance% = 18%

3 phase Y-Y Transformers: 20 MVA, 120Y/15Y KV, Reactance% = 12%

3 phase Y-Δ Transformers: 18 MVA, 120Y/15 ΔKV Reactance% = 12%

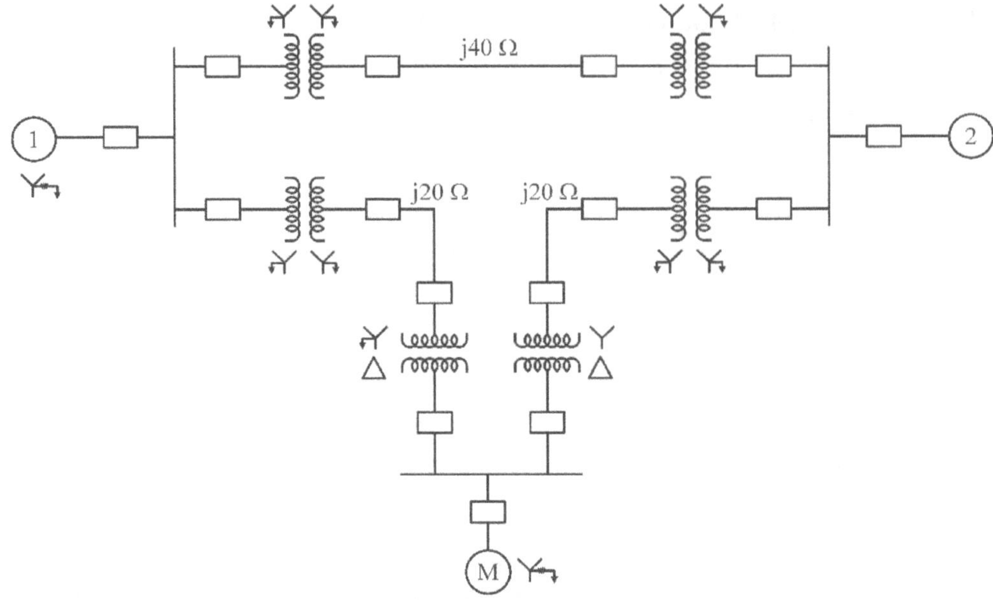

Fig.1.33

Solution:

Generator 1:

Zp.u.$_{given}$ = 0.2, Base KV$_{given}$ = 15KV, Base KV$_{Actual}$ = 15KV

Base MVA$_{Actual}$ = 40 KVA, Base MVA$_{given}$ = 25MVA

Zp.u.$_{new}$ = 0.2 × (15/15)^2 × (40/25000) = 0.00032pu

Transformer 1(Y-Y):

Zp.u.$_{given}$ = 0.12, Base KV$_{given}$ = 15 KV, Base KV $_{Actual}$ = 15KV,

Base MVA $_{Actual}$ = 40 KVA, Base MVA$_{given}$ = 20MVA

Zp.u.$_{new}$ = 0.12 × (15/15)^2 × (40/20000) = 0.00024p.u.

Transmission Lines:

$$\text{Per unit Impedance} = \frac{\text{Actual impedance}}{\text{Base impedance}}$$

$$\text{Base Impedance} = (\text{Base KV})^2/\text{Base MVA}$$

$$= \frac{(120)^2 \times 1000}{40}$$

$$= 360000$$

Per unit Impedance = 40/360000 = 0.000111 p.u.

Per unit Impedance on Other Transmission Line = 20/360000 = 0.000055556 p.u.

Generator 2:

Zp.u.$_{given}$ = 0.18, Base KV$_{given}$ = 16 KV, Base KV$_{Actual}$ = 15KV,

Base MVA$_{Actual}$ = 40KVA, Base MVA$_{given}$ = 28MVA

Zp.u.$_{new}$ = 0.18 × (16/15)^2 × (40/28000)

$\qquad$ = 0.00029257 p.u.

Transformer 2: (Y – Δ)

Zp.u.$_{given}$ = 0.12, Base KV$_{given}$ = 15KV, Base KV$_{Actual}$ = 15KV,

Base MVA $_{Actual}$ = 40 KVA, Base MVA$_{given}$ = 18MVA

Zp.u.$_{new}$ = 0.12 × (15/15)^2× (40/18000) = 0.000266 p.u.

Synchronous Motor 3:

Zp.u.$_{given}$ = 0.18, Base KV$_{given}$ = 14.8 KV, Base KV$_{Actual}$ = 15KV,

Base MVA $_{Actual}$ = 40KVA, Base MVA$_{given}$ = 33MVA

Zp.u.$_{new}$ = 0.18 × (14.8/15)^2 × (40/33000) = 0.00021204pu

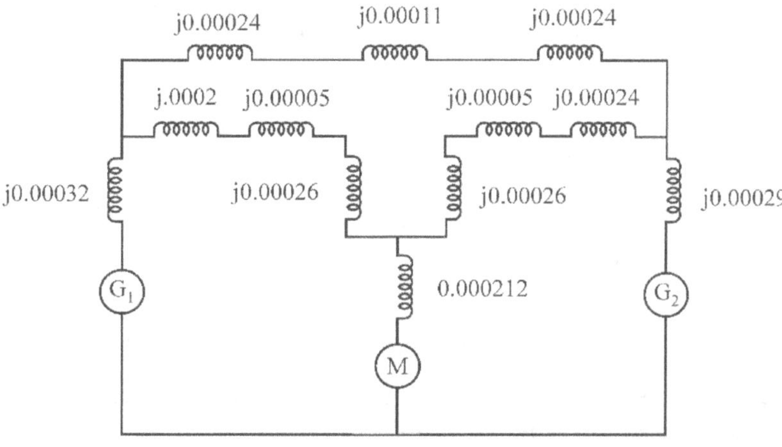

Fig.1.34

Example 21: A transformer having single phase is rated as 110/440V, 3KVA. L.T. side of the transformer has a leakage reactance of 0.05Ω. Calculate reactance in p.u.

Solution:

Actual leakage reactance = 0.05Ω

$$\text{Base impedance} = \frac{(\text{Base KV})^2}{\text{Base MVA}}$$

$$\text{Base impedance in L.T. side} = \frac{\left(110\times10^{-3}\right)^2 \times 1000}{3}$$

$$= 4.033 \ \Omega$$

$$\text{Per unit impedance of the circuit element} = \frac{\text{Actual impedance}}{\text{Base impedance}}$$

$$\text{Per unit impedance} = \frac{0.05}{4.033} = 0.01239 \text{p.u.}$$

Example 22: Below diagram shows a two machine system. The ratings are as follows:

Synchronous generator: 25MVA, 11KV, $X'' = 0.17$p.u.

Synchronous motors: 20MVA, 11KV, $X'' = 0.17$p.u.

Transformer T1:   30MVA, 12.5Δ/132YKV, $X = 0.12$p.u.

Transformer T2:   25MVA, 132Y/11ΔKV, $X = 0.12$p.u.

Line: $300 + j400$

Static load: 4MVA, 0.7 Power factor lagging.

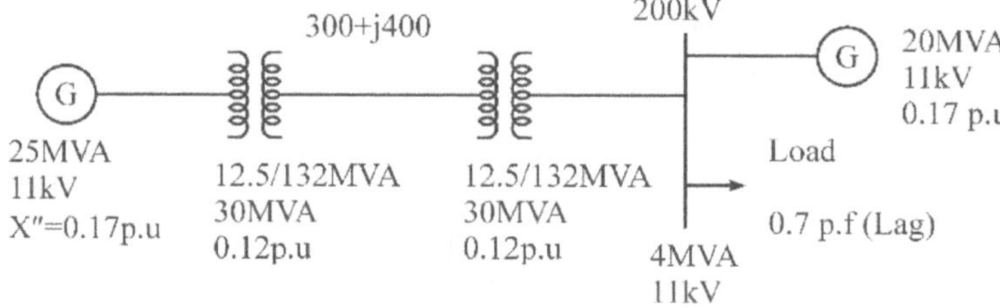

Fig.1.35

Draw the impedance diagram for the system. Choose base voltage of 132KV for the transmission line and a base volt ampere of 30MVA.

Solution:

Generator:

$$\text{Base voltage at Generator side} = \frac{125}{132}\times132 = 12.5\text{KV}$$

$$X'' = j0.17\times\frac{30}{25}\times\left(\frac{11}{12.5}\right)^2 = j0.157\text{p.u.}$$

Transformer 1:

Here we are taking base quantities at H.T. side of transformer.

$$X = j0.12 \times \frac{30}{30} \times \left(\frac{132}{132}\right)^2$$

$$= j0.12 \text{p.u.}$$

Transformer 2:

$$X'' = j0.12 \times \frac{30}{25} \times \left(\frac{132}{132}\right)^2$$

$$= j0.144 \text{p.u.}$$

Motor:

Base voltage at motor side $= \dfrac{11}{132} \times 132 = 11 \text{ KV}$

$$X'' = j0.17 \times \frac{30}{20} \times \left(\frac{11}{11}\right)^2$$

$$= j0.255 \text{p.u.}$$

Transmission line:

$$\text{Base impedance} = \frac{(\text{Base KV})^2}{\text{Base MVA}} = \frac{(132)^2}{30} = 580.8\Omega$$

$$Z_{\text{p.u.}} = \frac{\text{Actual impedance}}{\text{Base impedance}} = \frac{300 + j40}{580.8} = 0.5165 + j0.688$$

For a load of 5 MVA, 0.7 power factor lagging.

$$\cos\Phi = 0.7$$

$$\Phi = \cos^{-1}(0.7) = 45.57°$$

$$P = S\cos\Phi = 4 \times 0.7 = 2.8 \text{MW}$$

$$Q = \sin\Phi = 4 \times \sin(45.57°) = 2.86 \text{MVAr}$$

If load is represented as a series impedance,

$$R_{\text{p.u.}} = V_{\text{p.u.}}^2 \times S_B \times \frac{P}{P^2 + Q^2}$$

$$X_{\text{p.u.}} = V_{\text{p.u.}}^2 \times S_B \times \frac{Q}{P^2 + Q^2}$$

Here $\quad V_{\text{p.u.}} = \dfrac{11}{11} = \text{p.u.}$

$$S_b = 30 \text{ MVA}$$

$$R_{p.u.} = 1^2 \times 30 \times \frac{2.8}{2.8^2 + 2.86^2} = 5.24 \text{ p.u.}$$

$$X_{p.u.} = 1^2 \times 30 \times \frac{2.8}{2.8 + 2.86^2} = j5.35 \text{ p.u.}$$

Per unit impedance Diagram:

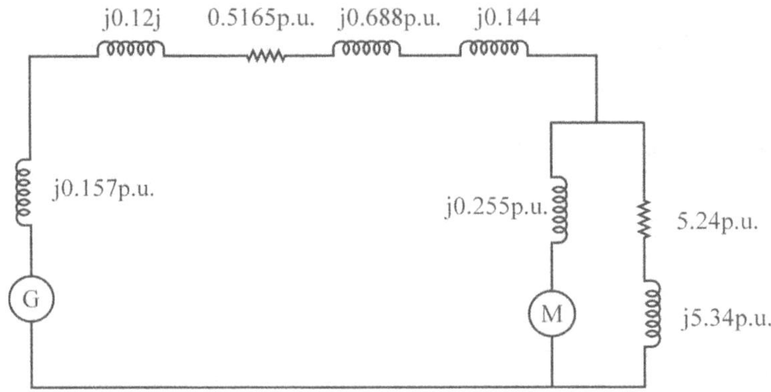

Fig.1.36

Example 23: Ratings of 3 generators are given below

G_1: 75MVA, 30KV, 9% Reactance

G_2: 125MVA, 31KV, 10% Reactance

G_3: 100MVA, 32KV, 7% Reactance

Find generator's new reactance with base values 150MVA and 40KV.

Solution:

$$Z_{p.u.\, new} = Z_{p.u.\, old} \times \left(\frac{\text{Base KV}_{given}}{\text{Base KV}_{Actual}} \right)^2 \times \left(\frac{\text{Base MVA}_{Actual}}{\text{Base MVA}_{given}} \right) \times \left(\frac{\text{Base MVA}_{Actual}}{\text{Base MVA}_{given}} \right)$$

G_1: $X_{G1} = 0.09 \times (30/40)^2 \times (150/75) = 0.1012$ p.u.

G_2: $X_{G2} = 0.1 \times (31/40)^2 \times (150/125) = 0.0721$ p.u.

G_3: $X_{G3} = 0.07 \times (32/40)^2 \times (150/100) = 0.0672$ p.u.

Example 24: Determine the generator voltage for given figure:

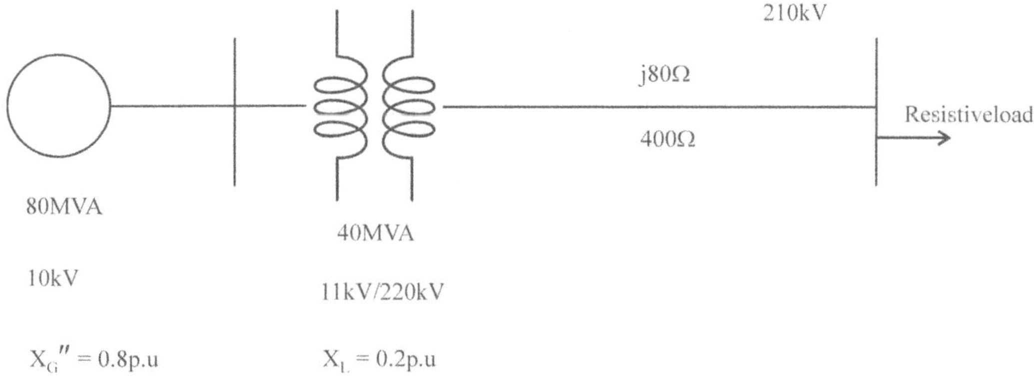

Fig.1.37

Solution: Assume 80MVA as Base power and 210KV as Base voltage in transmission line.

Generator:

Base KV$_{new}$ towards generator side = Transformer ratio× Base KV towards transmission line

Base KV$_{new}$ = (11/220) × 210 = 10.5

$$X_G'' = 0.8 \times (10/10.5)^2 \times (80/80)$$

$$X_G'' = 0.7256 \text{ p.u.}$$

Transformer:

$$X_T = 0.2 \times (220/210)^2 \times (80/40) = 0.4390 \text{ p.u.}$$

Transmission line: Base impedance $= \dfrac{(\text{Base KV})^2}{\text{Base MVA}} = \dfrac{210^2}{80}$

$$Z_{Base} = 551.25 \ \Omega$$

Per unit impedance = Actual impedance/Base impedance

$$= 80/551.25 = 0.1451 \text{ p.u.}$$

Resistive load 400Ω:

$$R_{p.u.} = 400/551.25 = 0.7256 \text{ p.u.}$$

Impedance diagram:

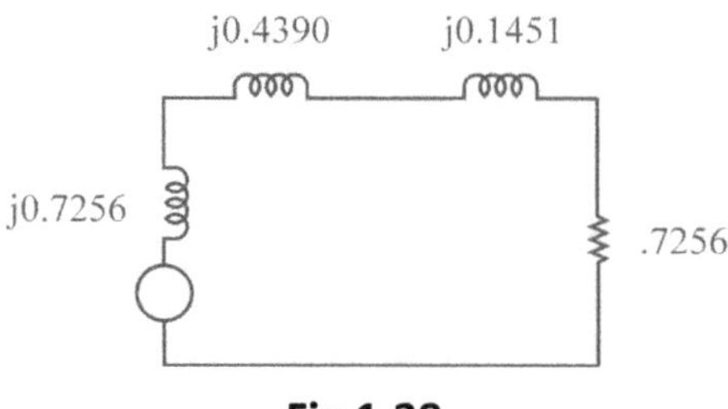

Fig.1.38

$V_{p.u.} = 210/210 = 1$ p.u.

$R_{p.u.} = 0.7256$ p.u.

Now, $I_{p.u.} = V_{p.u.}/R_{p.u.} = 1/0.7256 = 1.3781$ p.u.

Let, $I_{p.u.}$ is reference phasor

$I_{p.u.} = 1.3781 + j0 = 1.3781$ with 0^0 angle

According to KVL, the V_{drop} in network $= I_{p.u.}\{R_{p.u.} + j\,(X_G + X_T + X_{TL})\}$

$$= 1.3781\{0.7256 + j\,(0.7256 + 0.4390 + 0.1451)\}$$

$$= 0.9999 + j1.8049$$

$$= 2.0634 \text{ p.u.}$$

Actual generator voltage:

$V_G = V_{p.u.} \times$ Base Voltage

$V_G = 2.0634 \times 10.5 = 21.6657$ KV (line to line)

Example 25: The one line diagram of an unloaded power system is shown in Fig.1.39. Reactance of the two section of transmission line are shown in the diagram. The generators and transformers ratings are as follows:

Generator 1: 30MVA, 13.8KV, $X'' = 0.1$pu

Generator 2: 20MVA, 18KV, $X'' = 0.1$pu

Generator 3: 30MVA, 20KV, $X'' = 0.1$p.u.

Transformer 1: 30MVA, 13.8/200 KV, $X'' = 0.2$p.u.

Transformer 2: Single phase units each rated 50MVA, 115.4/18KV, $X'' = 0.2$p.u.

Transformer 3: 40MVA, 200Y/22YKV, $X'' = 0.2$p.u.

Draw the impedance diagram with all reactance marked in p.u. and with letters to indicate points corresponding to one-line diagram. Choose a base of 60MVA, 13.8KV in the circuit of generator 1.

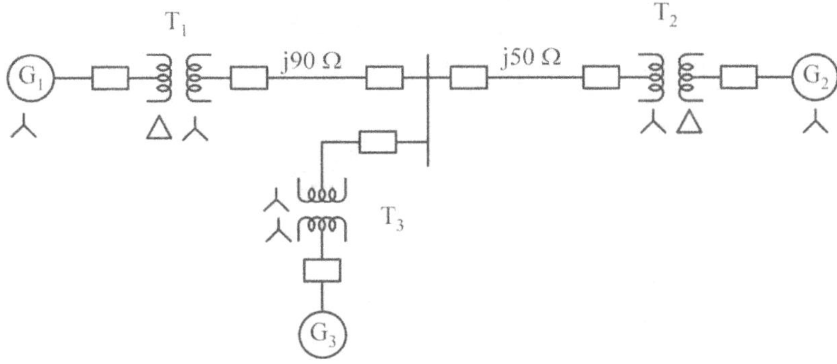

Fig.1.39

Solution:

Generator 1: 30MVA, 13.8KV, $X'' = 0.1$pu

Base MVA = 60MVA, Base KV = 13.8KV

$$Z_{p.u\ new} = Z_{p.u\ given} \times \left(\frac{\text{Base KV}_{given}}{\text{Base KV}_{new}}\right)^2 \times \left(\frac{\text{Base KMVA}_{new}}{\text{BaseMVA}_{given}}\right)$$

$$X_{G1} = 0.1 \times \left(\frac{13.8}{13.8}\right)^2 \times \left(\frac{60}{30}\right)$$

$$= 0.2 \text{ p.u.}$$

Transmission Line 1:

Base voltage on the transmission line using the transformation ratio = 200 KV.

$$\text{Base impedance} = \frac{(\text{Base KV})^2}{\text{Base MVA}}$$

$$\text{Base impedance} = \frac{(200)^2}{60} = 666.67\Omega$$

$$\text{Per unit impedance} = \frac{\text{Actual impedance}}{\text{Base impedance}}$$

$$= \frac{90}{666.67} = 0.134 \text{p.u}$$

Transmission Line 2:

$$\text{Per unit impedance} = \frac{\text{Actual impedance}}{\text{Base impedance}}$$

$$= \frac{50}{666.67} = 0.0749 \text{ p.u}$$

Transformer 1: Base MVA = 60 MVA, Base KV = 13.8 KV

Base MVA given = 30, Base KV$_{given}$ 13.8KV, $X'' = 0.2$ p.u

$$Z_{p.u\ new} = Z_{p.u\ given} \times \left(\frac{\left(Base\ KV_{given} \right)}{Base\ KV_{new}} \right)^2 \times \left(\frac{Base\ MVA_{new}}{Base\ MVA_{given}} \right)$$

$$X_{T1} = 0.2 \times \left(\frac{13.8}{13.8} \right)^2 \times \left(\frac{60}{30} \right) = 0.4\ p.u.$$

Transformer 3:

Base MVA = 60 MVA, Base KV = $200 \times \left(\frac{22}{200} \right) = 22$ KV

Given: 40 MVA, 220Y/22Y KV, $X'' = 0.2$p.u

$$Z_{p.u\ new} = Z_{p.u\ given} \times \left(\frac{\left(BaseKVgiven \right)}{BaseKVnew} \right)^2 \times \left(\frac{BaseMVAnew}{BaseMVAgiven} \right)$$

$$X_{T3} = 0.2 \times \left(\frac{22}{22} \right)^2 \times \left(\frac{60}{40} \right)$$

$$= 0.3\ p.u.$$

Transformer 2:

Voltage rating of transformer when they are put to form 3-Φ transformer,

$$= \frac{\sqrt{3}}{18} \times 115.4\ KV = \frac{200}{18} KV$$

$$Base\ KV = \frac{18}{200} \times 200 = 18KV$$

$$Z_{p.u\ new} = Z_{p.u\ given} \times \left(\frac{\left(BaseKVgiven \right)}{BaseKVnew} \right)^2 \times \left(\frac{BaseMVAnew}{BaseMVAgiven} \right)$$

$$X_{T2} = 0.2 \times \left(\frac{18}{18} \right)^2 \times \left(\frac{60}{50} \right) = 0.24\ p.u.$$

Generator 2:

Given: 20MVA, 18KV, $X'' = 0.1$p.u

Base MVA = 60 MVA, Base KV = 18 KV

$$Z_{p.u\ new} = Z_{p.u\ given} \times \left(\frac{(BaseKVgiven)}{BaseKVnew}\right)^2 \times \left(\frac{BaseMVAnew}{BaseMVAgiven}\right)$$

$$X_{G2} = 0.1 \times \left(\frac{18}{18}\right)^2 \times \left(\frac{60}{20}\right) = 0.3\ p.u$$

Generator 3:

Given: 20 MVA, 20KV, $X'' = 0.1$p.u.

Base MVA = 60 MVA, Base KV = 22 KV

$$Z_{p.u\ new} = Z_{p.u\ given} \times \left(\frac{(BaseKVgiven)}{BaseKVnew}\right)^2 \times \left(\frac{BaseMVAnew}{BaseMVAgiven}\right)$$

$$X_{G3} = 0.1 \times \left(\frac{20}{22}\right)^2 \times \left(\frac{60}{20}\right) = 0.247\ p.u$$

Impedance Diagram:

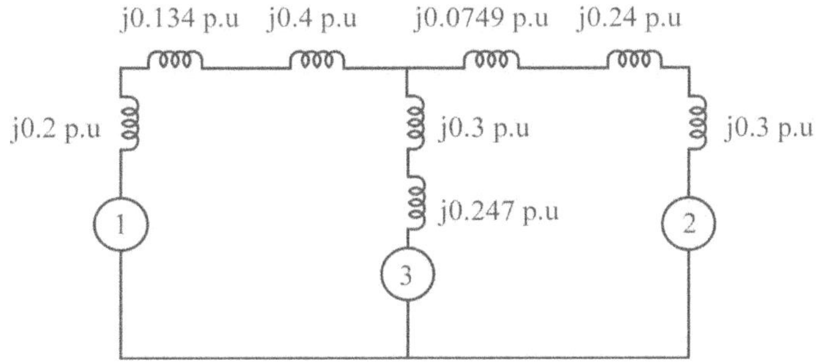

Fig.1.40

Impedance Diagram:

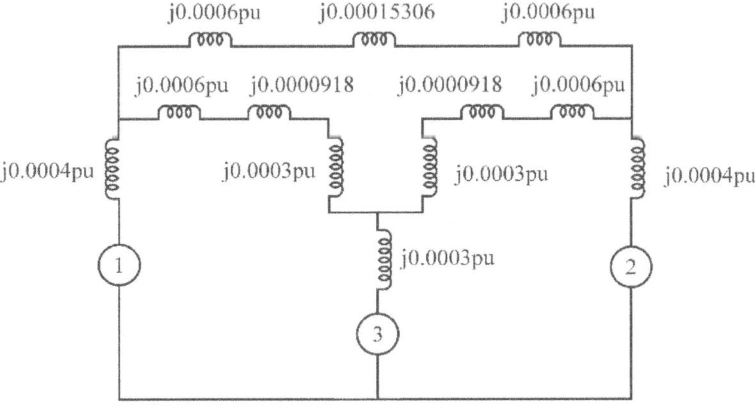

Fig. 1.41

Example 26: A 3 bus system is given in fig.1.41. The ratings of various components are listed below.

Generator 1: 70 MVA, 15.8 KV, $X'' = 017$ p.u.

Generator 2: 60 MVA, 15.2 KV, $X'' = 0.4$ p.u.

Generator 3: 50 MVA, 13 KV, $X'' = 0.35$ p.u.

Transformer 1: 55MVA, 13KV/120KV Y/Y, X = 0.2 p.u.

Transformer 2: 35 MVA, 14.5 KV/125 KV Y/Y, X = 0.16 p.u.

Transformer 3: 50 MVA, 14.5 кV/125 KV Y/Y, X = 0.2 p.u.

The line impedance are shown in figure. Determine the reactance base on 70 MVA and 15.8 KV as base quantities in generator 1.

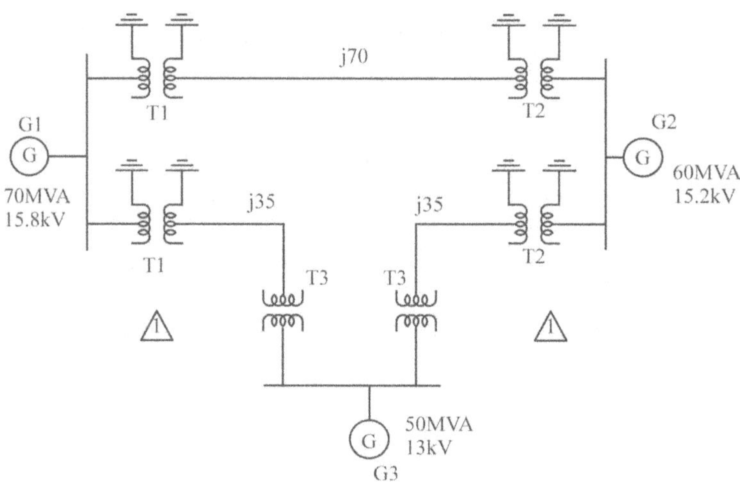

Fig.1.42

Solution: Base MVA = 70, Base KV = 15.8

Generator 1:
Per unit reactance = $0.17 \times (15.8/15.8)^2 \times 70/70$

$$= 0.17 \text{ p.u.}$$

Generator 2:

Base voltage = $145.8 \times 14.5/125 = 16.9$ KV
60 MVA, 15.2 KV, $X'' = 0.4$ p.u.

Base value 70 MVA, 16.9 KV

$X_{G2} = 0.4 \times (15.2/16.9)^2 \times 70/60$

$$= 0.3774 \text{ p.u.}$$

Generator 3: Base voltage on the bus = $145.8 \times 14.5/125 = 16.9$ kV

50MVA, 13KV, $X'' = 0.35$ p.u.

Base value 70MVA, 16.9 KV

$X_{G3} = 0.35 \times (13/16.9)^2 \times 70/50 = 0.2899$ p.u.

Transformer 1: 55 MVA, 13 KV, X = 0.2 p.u

Base value 70 MVA and 15.8 KV

$X_{T1} = 0.1 \times (13/15.8)^2 \times 70/55$

$\qquad = 0.0861$ p.u

Transformer 2: 35 MVA, 14.5 KV, X = 0.16 p.u.

Base value = $145.8 \times 14.5/125 = 16.9$ KV

$X_{T2} = 0.16 \times (14.5/16.9)^2 \times 70/35$

$\qquad = 0.2355$ p.u

Transformer 3: 50MVA, 12.5 KV, X = 0.2 p.u.

Base value = $16.9 \times 125/14.5$

$\qquad = 145$ KV

$X_{T3} = 0.2 \times (125/145)^2 \times 70/50$

$\qquad = 0.2080$ p.u.

Transmission line 1: base voltage along the transmission line whose impedance is

$j70\Omega = 15.8 \times 120/13$

$\qquad = 145.8$ KV

$$\text{Base impedance} = \frac{(\text{Base KV})^2}{\text{Base MVA}}$$

$\qquad = 145^2/70 \; = 303.68 \; \Omega$

$$\text{Per unit impedance} = \frac{\text{Actual impedance}}{\text{Base impedance}}$$

$\qquad = 70/303.68 = 0.2305$ p.u

Transmission line 2: Base KV = 145.8, Base MVA = 70

Base impedance = j 37.5/303.68

$\qquad = $ j 0.1152 p.u

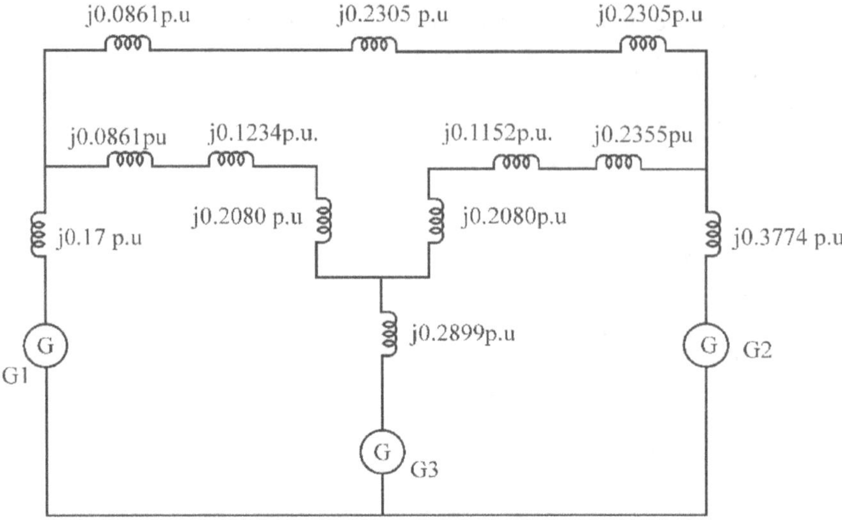

Fig.1.43

Example 27: A 25 MVA, 9.5 KV, 3-Φ generator has subtransient reactance of 20%. It is connected through a Δ-Y transformer to a high voltage transmission line having a total series reactance of 70 Ω. At the load end of the line is a Y-Y step down transformer. Both transformer banks are composed of 1-Φ transformers connected for 3-Φ operation. Each of the two transformers comprising each bank is rated 7682 KVA, 10/100 KV with reactance of 10%. The load represented as impedance is drawing 12 MVA at 14.5 KV at 80% power factor lagging. Draw the impedance diagram showing all impedance in per unit. Choose a base of 12 MVA, 14.5 KV in the load circuit.

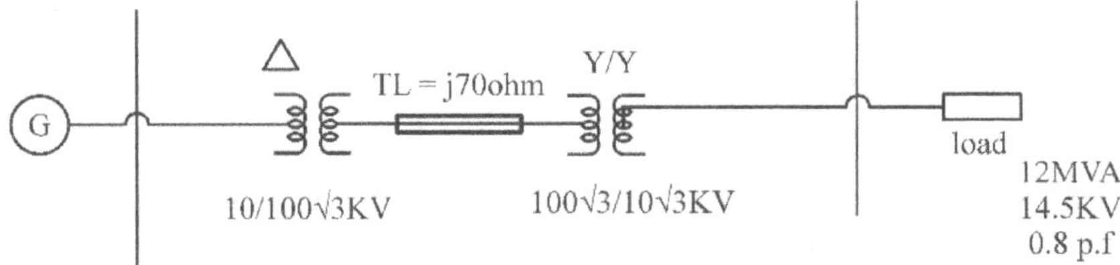

Fig.1.44

Solution:

Load:

MVA = 12, KV = 14.5

Load impedance = 0.8 + j o.6 p.u.

Total MVA rating of transformer 1 and 2 separately = 3 × 7628/1000

$$= 23 \text{ MVA}$$

Base KV on the transmission line side $= 14.5 \times \dfrac{100\sqrt{3}}{10\sqrt{3}}$

$$= 145 \text{ KV}$$

Base impedance $= (\text{Base KV})^2/\text{Base MVA}$

Base impedance in the transmission line $= 145^2/12$

$$= 1752.08 \ \Omega$$

Per unit impedance $=$ Actual impedance/Base impedance

$$= 70/1752.08 = 0.0399 \text{ p.u.}$$

Transformer 1: Base KV on the generator side is given by

$$= 145 \times 10/100 \sqrt{3} = 8.3 \text{ KV}$$

$$X_{T1} = 0.1 \times (10/8.3)^2 \times (12/23)$$

$$= 0.075 \text{ p.u}$$

Generator 1: given 25 MVA, 9.5KV, $X'' = 0.2$ p.u.

Base value $= 12$MVA, 8.3 KV

$$X'' = 0.2 \times (9.5/8.3)^2 \times (12/25)$$

$$= j\, 0.1257 \text{ p.u}$$

Transformer 2: given 23MVA, $100 \sqrt{3}$ KV, $X'' = 0.1$ p.u.

Base value $= 12$ MVA, 145KV

$$X_{T2} = 0.1 \times (100 \sqrt{3} /145)^2 \times 12/23$$

$$= 0.0743 \text{ p.u}$$

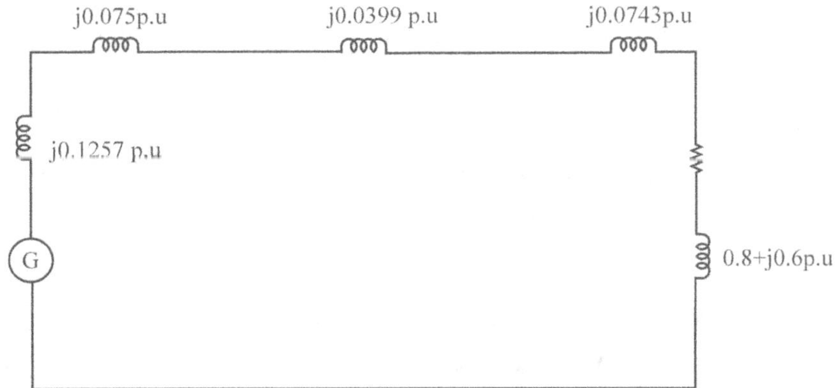

Fig.1.45

Example 28: A 200 MVA,44KV, 3-ϕ generator has a subtransient reactance of 25%. The generator is connected to three motors through a transmission line and two Transformers. Each motor have rated 40MVA with 30% subtransient reactance. The 3-ϕ transformers are rated at 110MVA,32KV/110KV with leakage reactance 9%. The line has a reactance of 50 ohms. Selecting the generator rating as the base quantities in the generator circuit, determine the base quantities in other parts of the system and evaluate the corresponding p.u. values.

Solution:

The base voltage in the transmission line = 44 × 110/32 = 151.25KV

In the motor circuit, the base voltage = 151.25 × 32/110 = 44KV

Base impedance in the transmission line = (Base KV)^2/base MVA

$$= (151.25)^2/200 = 114.382 \ \Omega$$

p.u. impedance = j50/j114.382 = 0.4371 p.u.

Motor 1:

40 MVA, 44KV, $X'' = 0.3$

$$X_1'' = 0.3 \times (40/44)^2 \times 200/40 = 0.4371 \text{ p.u.}$$

Motor 2:

30MVA, 40KV, $X'' = 0.3$

$$X_2'' = 0.3 \times (40/44)^2 \times 200/30 = 1.6528 \text{ p.u.}$$

Motor 3:

60MVA, 40KV, $X'' = 0.3$

$$X_3'' = 0.3 \times (40/44)^2 \times 200/60 = 0.8264 \text{ p.u.}$$

Transformer:

110MVA, 32KV, $X'' = 0.09$

$$X_T'' = 0.09 \times (32/44)^2 \times 200/110$$

$$= 0.0865 \text{ p.u.}$$

Example 29: Fig.1.45 shows a two machine system. The ratings are as follows:

Synchronous generator: 30MVA, 11KV, $X'' = 0.25$ p.u.

Synchronous motor: 25MVA, 11KV, $X'' = 0.25$ p.u.

Transformer T_1 = 35MVA, 13.5Δ/132Y KV, X = 0.2 p.u

Transformer T_2 = 30MVA, 132Y/11ΔKV, X = 0.2 p.u

Line: 100 + j400Ω

Static load: 10MVA, 0.75 power factor lagging.

Draw the impedance diagram for the system. Choose a base voltage of 132KV for the transmission line and a base volt ampere of 30MVA.

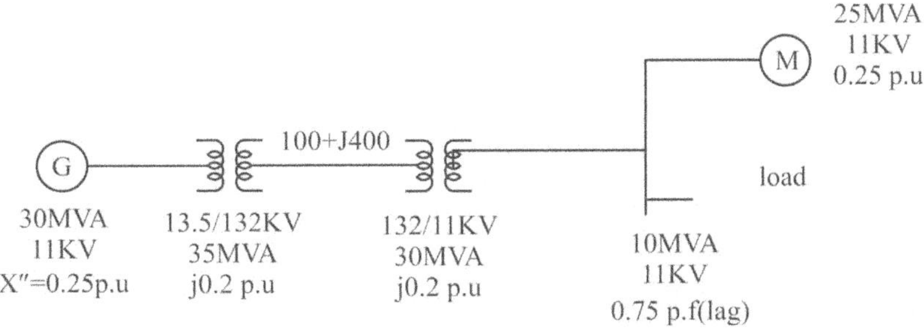

Fig.1.46

Solution:

Generator:

$$X'' = j0.25 \times 30/30 \times (11/13.5)^2 = j0.1659 \text{ p.u.}$$

Transformer 1:

$$X = j0.2 \times 30/35 \times 132/132 = j0.1714 \text{ p.u.}$$

Transformer 2:

$$X'' = j0.2 \times 30/30 \times (132/132)^2 = j0.2 \text{ p.u.}$$

Motor:

$$X'' = 0.25 \times 30/25 \times (11/11)^2 = 0.3 \text{ p.u.}$$

Transmission line:

$$\text{Base impedance} = \frac{\left(\text{Base KV}\right)^2}{\text{Base MVA}}$$

$$= 132^2/30 = 580.8\,\Omega$$

$$Z_{p.u.} = \frac{\text{Actual impedance}}{\text{Base impedance}}$$

$$= 100 + j400/580.8$$

$$= 0.172 + j0.688$$

For a load of 10MVA, 0.75 power factor lagging,

$$P = S\cos\phi = 10 \times 0.75 = 7.5\text{MW}.$$

$$Q = S\sin\phi = 10 \times 0.66 = 6.6\text{MVAr}$$

If load is represented as a series impedance,

$$R_{p.u} = \frac{V_{p.u}^2 . S_b . P}{P^2 + Q^2} \quad \text{and} \quad X_{p.u} = \frac{V_{p.u}^2 . S_b . Q}{P^2 + Q^2}$$

Here, $V_{p.u.} = 11/11 = 1 \text{p.u}$

$S_b = 30\text{MVA}$

Therefore, $R_{p.u} = \dfrac{(1.0)^2 \times 30 \times 7.5}{(7.5)^2 + (6.6)^2} = 2.25 \text{ p.u.}$

$$X_{p.u} = \frac{(1.0)^2 \times 30 \times 6.6}{(7.5)^2 + (6.6)^2} = 1.983 \text{ p.u.}$$

P.U. impedance diagram:

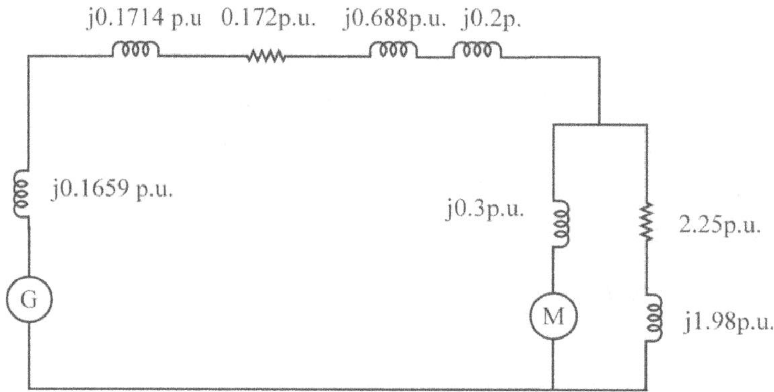

Fig.1.47

Example 30: Three generators are rated as follows

Generator 1: 120MVA, 32KV, reactance 9%.

Generator 2: 140MVA, 30KV, reactance 7%.

Generator 3: 160MVA, 29KV, reactance 10%.

Determine the reactance of the generators corresponding to base values of 250MVA and 40KV.

Solution:

Generator 1:

120MVA, 32KV, $X'' = 0.09$

$$Z_{p.u.new} = Z_{p.u.given} \times \left(\frac{\text{Base KVgiven}}{\text{Base KVnew}}\right)^2 \times \left(\frac{\text{Base KMVAnew}}{\text{Base MVAgiven}}\right)$$

$$X'' = 0.09 \times (32/40)^2 \times 250/120$$

$$= 0.09 \times 0.64 \times 2.083$$

$$= 0.1199 \text{ p.u.}$$

Generator 2:

$$140\text{MVA, } 30\text{kV, } X'' = 0.07$$

$$Z_{p.u.new} = Z_{p.u.given} \times \left(\frac{\text{Base KVgiven}}{\text{Base KVnew}} \right)^2 \times \left(\frac{\text{Base KMVAnew}}{\text{Base MVAgiven}} \right)$$

$$X'' = 0.07 \times (30/40)^2 \times 250/140$$

$$= 0.07 \times 0.5625 \times 1.785$$

$$= 0.0702 \text{ p.u.}$$

Generator 3:

$$160\text{MVA, } 29\text{KV, } X'' = 0.1$$

$$Z_{p.u.new} = Z_{p.u.given} \times \left(\frac{\text{Base KVgiven}}{\text{Base KVnew}} \right)^2 \times \left(\frac{\text{Base KMVAnew}}{\text{Base MVAgiven}} \right)$$

$$X'' = 0.1 \times (29/40)^2 \times 250/160$$

$$= 0.1 \times 0.5256 \times 1.5625$$

$$= 0.0821_{p.u.}$$

Example 31: Determine the generator voltage for the network as shown in the figure below.

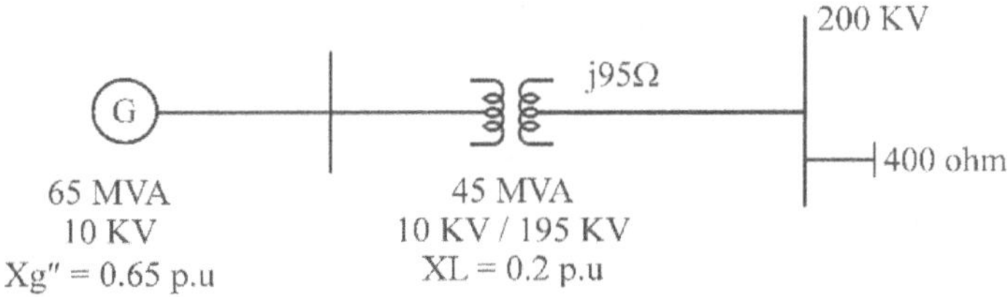

Fig.1.48

Solution:

Let us assume a Base value of the transmission line is 95MVA & 200KV.

Generator: Base voltage of the Generator: (10/195) × 200 kV = 10.2564kV

$$Z_{p.u.new} = Z_{p.u.given} \times \left(\frac{\text{Base KVgiven}}{\text{Base KVnew}} \right)^2 \times \left(\frac{\text{Base MVAnew}}{\text{Base MVAgiven}} \right)$$

$$X''_G = 0.65 \times 95/65 \times (10/10.2564)^2$$

$$= 0.65 \times 1.4615 \times 0.9506$$

$$= 0.90304 \text{p.u.}$$

Transformer:

$$X_T = 0.2 \times 95/45 \times (195/200)^2$$
$$= 0.2 \times 2.111 \times 0.9506$$
$$= 0.4013 \text{ p.u.}$$

Transmission Line:

$$\text{Base impedance} = \frac{(\text{Base KV})^2}{\text{Base MVA}}$$
$$= (200)^2/95$$
$$= 40000/95$$
$$= 421.05 \ \Omega$$

$$\text{Per unit impedance} = \frac{\text{Actual limpedance}}{\text{Base impedance}}$$
$$= 95/421.05$$
$$= 0.225 \text{p.u.}$$

Resistive load of 400 Ω:

$$R_{P.U} = 400/421.05$$
$$= 0.95 \text{p.u.}$$

Impedance diagram:

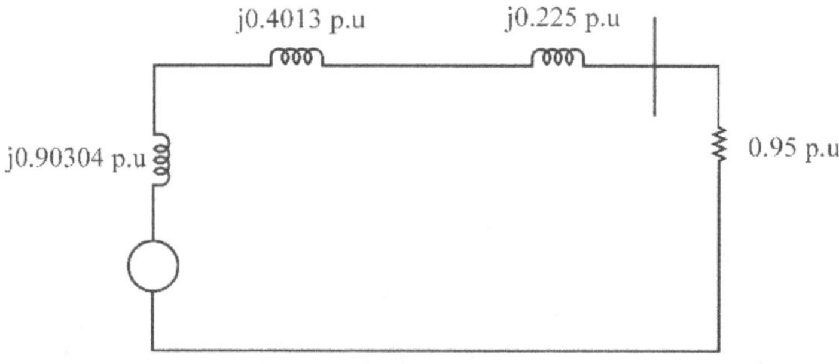

Fig.1.49

At the load bus,

$$V_{p.u} = 1 \text{p.u.}$$
$$R_{p.u.} = 0.95 \text{p.u.}$$

Therefore,

$$I_{p.u.} = V_{p.u.}/R_{p.u.}$$

=1/0.95

=1.052p.u.

Let us take $I_{p.u.}$ as the reference phasor.

$I_{p.u}$=1.052p.u+j0

=1.052 with angle 0^0

Applying KVL, the voltage drop in the network

$= I_{p.u}[R_{p.u} + j(X_G + X_t + X_L)]$

$= 1.052[0.95 + j(0.90304 + 0.4013 + 0.225)]$

$=1.052[0.95 + j1.52934]$

$= 1.052 \times 1.8003$

$= 1.8939p.u.$

Actual generator terminal voltage,

$V_G = V_{p.u} \times$ Base voltage at the generator terminals

$= 1.8939 \times 10.2564$

$= 19.425$ kV (Line to Line)

Example 32: A 3 bus system is given in figure. The ratings of the various components are listed below:

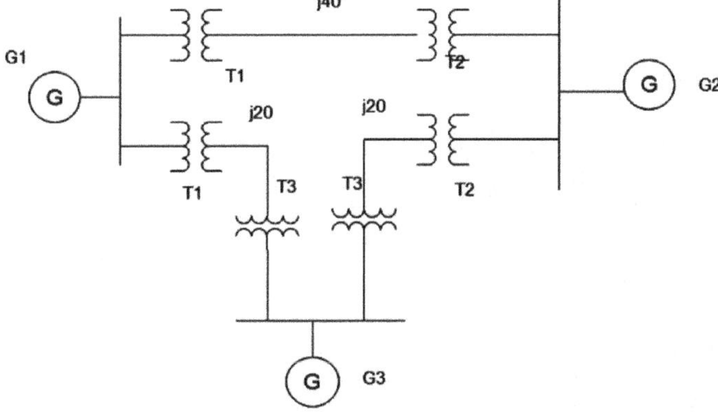

Fig.1.50

Generator 1: 40MVA, 12.8kV, $X" = 0.2_{p.u}$

Generator 2: 50MVA, 12.2Kv, $X" = 0.15_{p.u}$

Generator 3: 20MVA, 12Kv, $X" = 0.20_{p.u}$

Transformer 1: 40MVA, 12kV/100kV Y/Y,$X = 0.1_{p.u}$

Transformer 2: 20MVA, 13.5kV/125kV Y/Y,$X = 0.20_{p.u}$

Transformer 3: 35NVA, 10.5Kv/115kV Y/Y,$X = 0.2_{p.u}$

The line impedances are shown in figure 50. Determine the reactance diagram based on 40MVA and 12.8kV as base quantities in generator 1.

Solution:

Base MVA=40; BasekV =12.8

Generator 1:

$Z_{p.u.new} = Z_{p.u.given}$(Base kV $_{given}$/Base kV $_{new}$)^{2}x Base MVA$_{new}$/Base MVA$_{given}$

Per Unit Reactance = $0.2 \times (12.8/12.8)^2 \times (40/40)$

$$= 0.0156_{p.u}$$

Transmission Line 1:

Base voltage along the transmission line whose impedance is $j40\Omega = 12.8 \times 100/12 = 106kV$

Base impedance = (Base kV)^2/Base MVA

Base impedance = $106^2/40 = 280.9\Omega$

Per unit impedance = $40/280.9 = 0.1423_{p.u}$

Generator 2:

Base voltage = $106 \times 12.5/10.5 = 12$ kV

$$X_{G2} = 0.15 \times (12.2/12)^2 \times 40/50$$

$$= 4.4652 \text{ p.u}$$

Transformer 1:

40MVA, 12Kv, $X = 0.1_{p.u}$

Base values; 40MVA, 12.8Kv

$$X_{T2} = 0.1 \times (12/12.8)^2 \times 40/40$$

$$= 0.0878_{p.u}$$

Transformer 2:

20MVA, 11.5Kv, $X = 0.20_{p.u}$

Base Voltage = $106 \times 13.5/125 = 11kv$

$$X_{T2} = 0.1 \times (11.5/11)^2 \times 40/20$$

$$= 0.2185_{p.u}$$

Transformer 3:

Base Voltage on the H.T side of transformer 3 is 106kv

$$X_{T3} = 0.2 \times (115/108)^2 \times 40/35$$

$$= 0.25916 \text{ P.U}$$

Generator 3:

Base voltage on the bus = 106 × 13.5/125 = 11kv

20 MVA, 12 kV, X" = 0.20p.u

Base value: 40MVA, 11kv

$$X_{G3} = 0{,}20 \times (12/11)^2 \times 40/20$$

$$= 0.476 \text{ P.U}$$

Transmission line 2:

Base kV =106

Base MVA = 40

Base impedance = $106^2/40 = 280.9$

Per unit impedance = j20/280.9 = j0.0711

Reactance Diagram:

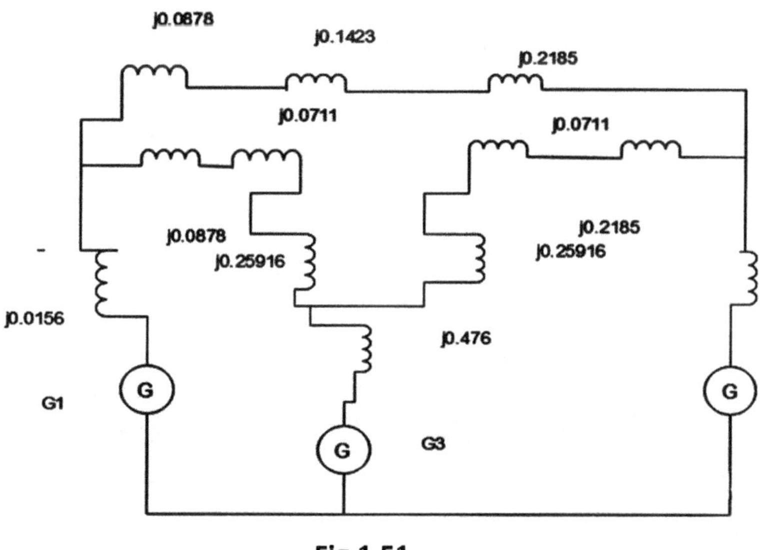

Fig.1.51

Example 33: Draw the impedance for the power system shown in fig. 1.49. Impedances are in p.u. Neglect resistances. Use a base of 45kVA, 148 KV, 20Ω line. The ratings of the generators, motors and transformers are:

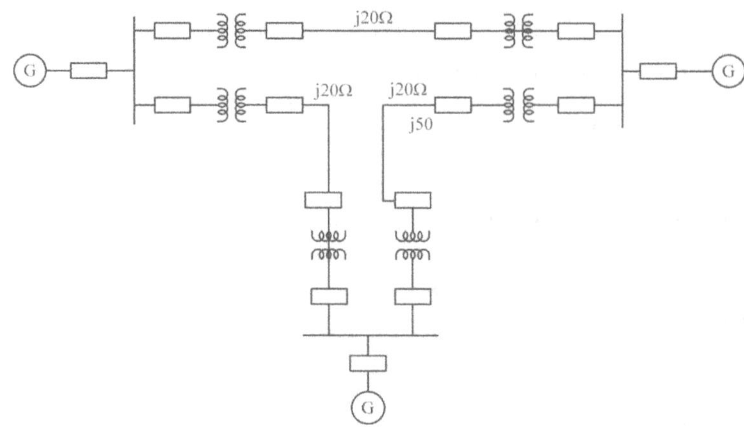

Fig. 1.52

Generator 1: 25MVA, 18KV, $X'' = 20\%$

Generator 2: 25MVA, 18KV, $X'' = 20\%$

Synchronous motor 3; 35 MVA, 12.8KV, $X'' = 20\%$

Three phase Y-Y Transformers: 10MVA, 128Y/20YKV, $X = 10\%$

Three phase Y-Δ Transformers: 10MVA, 128Y/12.8ΔKV, $X = 10\%$

Solution:

Generator 1:

Given: 25MVA, 18 KV, $X'' = 20\%$

Base values: 45 KVA

$$Z_{p.unew} = Z_{p.ugiven} \times \left(\frac{\text{Base KVgiven}}{\text{Base KVnew}} \right)^2 \times \left(\frac{\text{Base KMVAnew}}{\text{Base MVAgiven}} \right)$$

$X_{G1} = 0.2 \times (18/20)^2 \times (45/20 \times 10^3)$

$\quad = 0.000364$ P.U

Transformer 1:

Given: 25MVA, 20KV, $X = 10\%$

Base values; 45KVA, 20KV

$X_{T1} = 0.1 \times (45/20 \times 10^3) = 0.00025$ P.U

Transmission Lines:

$$\text{Base impedance} = \frac{\left(\text{Base KV} \right)^2}{\text{Base MVA}}$$

Base impedance = $148^2/40 \times 1000$

$$= 547600\Omega$$

$$\text{Per unit impedance} = \frac{\text{Actual impedance}}{\text{Base impedance}}$$

p.u impedance = 0.0000821 P.U

p.u impedance of the other transmission line = 20/547600

$$= 0.0000807 \text{ P.U}$$

Generator 2:

$$X_{G2} = X_{G1} = 0.0003645 \text{ P.U}$$

Y-Δ transformer

Given: 10MVA, 128Y/12.8 ΔKV, X = 10%

Base values; 45kVA, 128KV

$X = (0.1 \times 45/10 \times 10^3)$

$= 0.000045$ P.U

Motor:

Given: 35MVA, 12.8 KV, $X'' = 20\%$

Base values: 45 KVA, 128KV

$$X_m = 0.2 \times 45/35 \times 10^3$$

$$= 0.0000257 \text{ P.U}$$

Impedance diagram

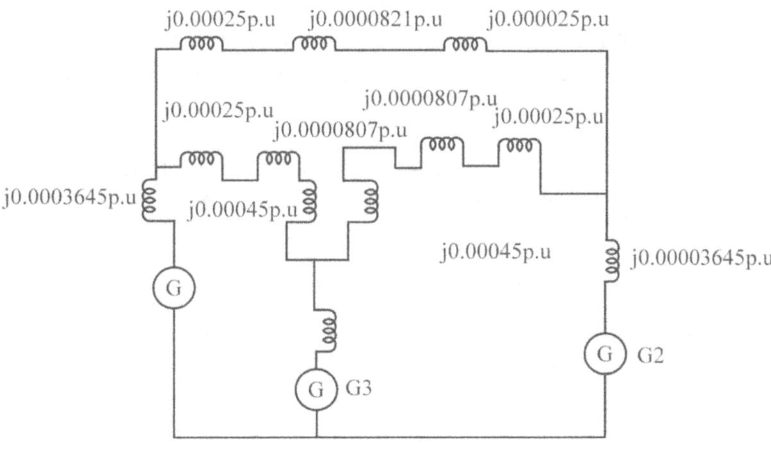

Fig.1.53

Exercise

Multiple Choice Questions

1. What is the value of base impedance for a 20kV, 100MVA, 3-ϕ System?

   a) 5 Ω b) 7 Ω c) 4 Ω d) 1 kΩ

2. In a system, base quantities are 40 kV and 200 MVA. What the value of p.u. admittance of j0.7S?

   a) j4S   b) j5.6S   c) j5.9S   d) j6.4S

3. What is the per unit synchronous reactance on the base value of 300 MVA and 30kV if a 200 MVA with 40 kV synchronous generator has 2p.u synchronous reactance.

   a) 5.33pu   b) 5.56pu   c) 5.79pu   d) 5.2pu

4. Which one is correct: Per unit impedance of a transformer is ______

   a) higher if calculated from primary side than from secondary side.

   b) both are same whether calculate from primary side or secondary side.

   c) always zero.

   d) always unity

5. A synchronous machine's p.u. value is 0.484. What will the p.u. value be if the base voltage is increased by 1.5 times?

   a) 0.266   b) 0.242   c) 0.220   d) 0.215

6. A synchronous generator is capable of operating at 21 kV, 50 MVA and 50Hz. The generator's base impedance is going to be

   a) 8.82 Ω b) 8.9 Ω c) 8.1Ω d) 8.25Ω

7. The synchronous reactance of an 22kV, 300MVA synchronous generator is 0.5 p.u. According to the basis values of 200MVA and 11kV, the p.u. synchronous reactance is

   a) 1.64   b) 1.33   c) 1.43   d) 1.54

8. For a 275MVA machine operating on its own base, the p.u. parameters are inertia M = 13 pu and reactance X = 9 pu. On a 70 MVA common base, the pu values for inertia and reactance will be

   a) 51.07, 2.29   b) 40, 1.9   c) 60, 3.4   d) 49,10

9. A circuit element has a per-unit impedance of 0.25. The new value of the per-unit impedance of the circuit element will be if the base kV is reduced by one third and the base MVA is reduced by half

   a) 1.161   b) 1.125   c) 1.191   d) 1.233

10. The per unit impedance value of an element is represented by X for a given base voltage and base volt ampere. What will this element's per-unit impedance value be if the voltage and volt-amp bases are both doubled?

    a) 2X   b) 4X   c) 0.5X   d) X

Answers

   1. (c),   2. (b),   3. (a),   4. (b),   5. (d),   6. (a),   7. (b),   8. (a),   9. (b),   10. (c)

Subjective Questions

1. What are the advantages of Per Unit system?

2. Three generators are rated as follows:

   Generator 1 – 120 MVA, 40KV, reactance 14%.

   Generator 2 - 170 MVA, 35KV, reactance 9%.

   Generator 3 – 140 MVA, 31KV, reactance 16%.

   Determine the reactance of the generator corresponding to base values of 250 MVA and 45 KV.

3. A 3 bus system is given in fig.1.51. The ratings of the various components are listed below:

   Generator 1: 50 MVA, 13.8 KV, $X'' = 0.4$ P.U
   Generator 2: 60 MVA, 12 KV, $X'' = 0.25$ P.U
   Generator 3: 40 MVA, 11 KV, $X'' = 0.30$ P.U
   Transformer 1: 50 MVA, 10 KV/100KV Y/Y, $X = 0.24$ P.U
   Transformer 2: 30 MVA, 11.5 KV/110KV Y/Y, $X = 0.30$ P.U
   Transformer 3: 35 MVA, 11.5 KV/110KV Y/Y, $X = 0.5$ P.U

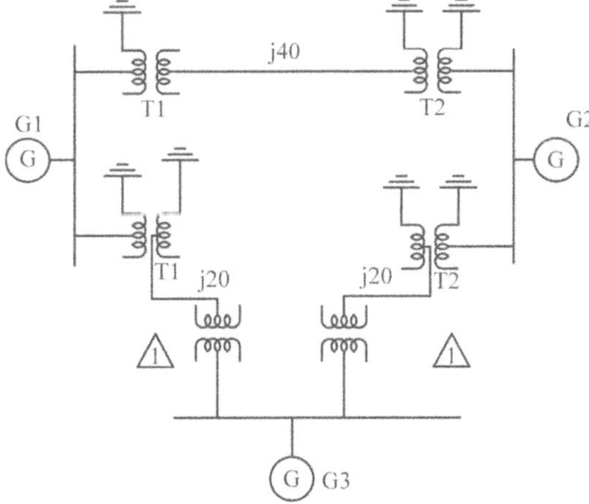

Fig.1.54

The line impedances are shown in the figure. Determine the reactance diagram based on 50 MVA and 13.8 KV as base quantities in Generator 1.

4. Two Generators rated at 15 MVA, 13 KV and 20 MVA, 13.2 KV are connected in parallel to a bus bar. They feed supply to two motors of input 10 MVA, and 12 MVA respectively. The Operating Voltage of motors is 11.2KV. Assuming base quantities as 25 MVA and 15KV, Draw the reactance diagram. The percentage Reactance for Generators is 13% and that for motors is 20%.

5. Ratings of 3 generators are given below:

   G_1: 125 MVA, 20KV, 9% Reactance

   G_2: 100 MVA, 35KV, 10% Reactance

   G_3: 150 MVA, 30KV, 7% Reactance

   Find generator's new reactance with base values 150MVA and 40KV.

6. A 300MVA, 44KV, 15%, 3-ϕ generator has a subtransient reactance of 20%.The generator is connected to three motors through a transmission line and two Transformers. The motors have rated inputs of 50MVA, 20MVA and 70MVA with 30% subtransient reactance. The 3-ϕ transformers are rated at 100 MVA, 32KV/110KV with leakage reactance 7%. The line has a reactance of 40 ohms. Selecting the generator rating as the base quantities in the generator circuit, determine the base quantities in other parts of the system and evaluate the corresponding p.u. values.

7. Three generators are rated as follows

   Generator 1 - 100MVA, 30KV, reactance 9%.

   Generator 2 - 150MVA, 35KV, reactance 11%.

   Generator 3 - 200MVA, 40KV, reactance 12%.

   Determine the reactance of the generators corresponding to Base values of 200MVA and 40KV.

Chapter 2

Introduction to Fault Analysis

2.0 Introduction

Almost all normal working mode of a Power System is balanced Three Phase System. A variety of unwanted but unavoidable occurrences can temporarily interrupt this condition. Such incidents can be stated as a fault.

A short-circuit fault happens due to two or more phase conductors which operate with different potential and touchdown each other through metallic or arc. Short-circuit faults in the power network may be categorized into Balanced or symmetrical three-phase faults and asymmetrical three-phase faults. Single line-to-ground faults, Line-to-line faults, Double line-to-ground faults are three different types of asymmetrical faults.

Electrical Machines such as Generator and Transformer may take balanced or symmetrical three-phase fault due to insulation breakdown between turns in the slot or winding and metallic structure which is connected to ground via mechanical structure. Due to heavy wind on the stormy day tree may be falling down to bare overhead lines as consequences of unbalanced faults are taking place. The other accidental phenomena such as supporting structure broken, birds fallen down to the bare overhead conductor and insulator failure may also cause the asymmetric faults.

The severity of a balanced fault is more in contrast to asymmetrical faults. Statistically, 75% of faults are single line to a ground type asymmetrical fault. The symmetrical fault is rarely occurring. However, the estimation of the maximum amount of fault current in the power network may be obtained by symmetrical fault analysis, which is an important element for relay coordination as well as for circuit breaker rating.

When fault occurs in a part of the power system, heavy current flows in that part of the circuit which may cause permanent damage to the equipments. Hence the faulty part should be isolated from the healthy part immediately on the occurrence of the fault. This can be achieved by providing protective relays which sense the faulty conditions and send signals to circuit breakers to open the circuit under faulty condition.

2.1 Transients due to Balanced Fault

Largely of the power system components are in an inductive nature which give rise to transients when there is a sudden change in current. Faults on the power network are accompanied by a sudden change in current which give rise to transient conditions in the power system.

Transients due to short circuit in transmission line.

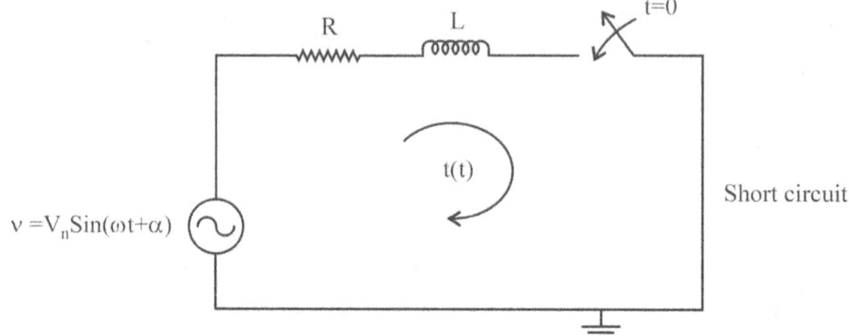

Fig.2.1

Z = Impedance of the transmission line.

$$= R + j\omega L = \sqrt{R^2 + \omega^2 L^2} < \left(\tan^{-1} \frac{\omega L}{R} \right) = 1Z1 < \theta$$

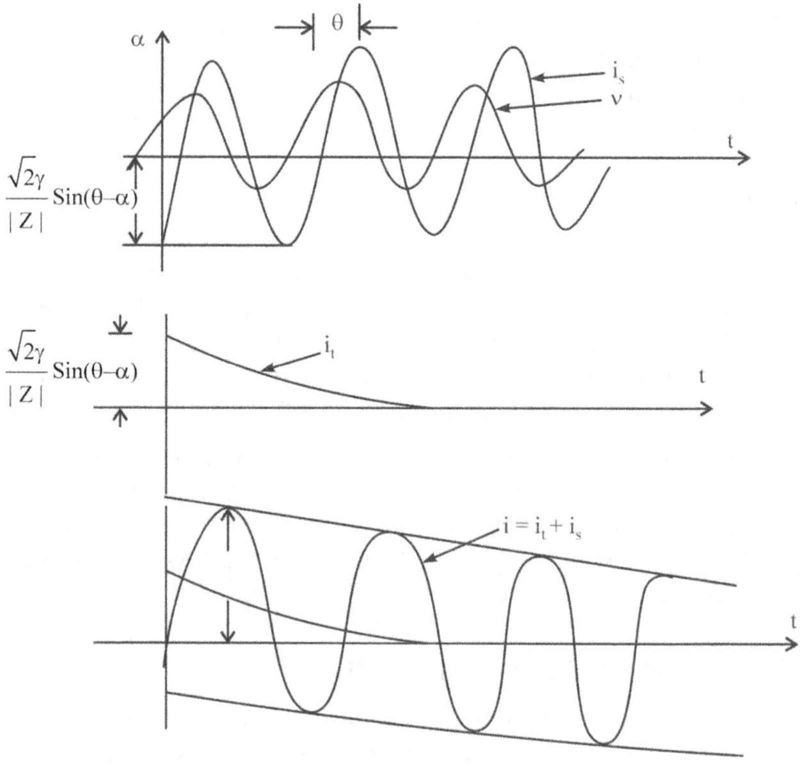

Fig.2.2

It is assumed that the short circuit has been simulated by closing the switch at t = 0 for the transmission line as shown in Fig.2.1. When the switch is closed at t = 0, fault current flows in the circuit. The differential equation for the short circuit which is shown in Fig.2.1 can be obtained by using KVL. The α controls the moments on voltage wave when a short circuit happens.

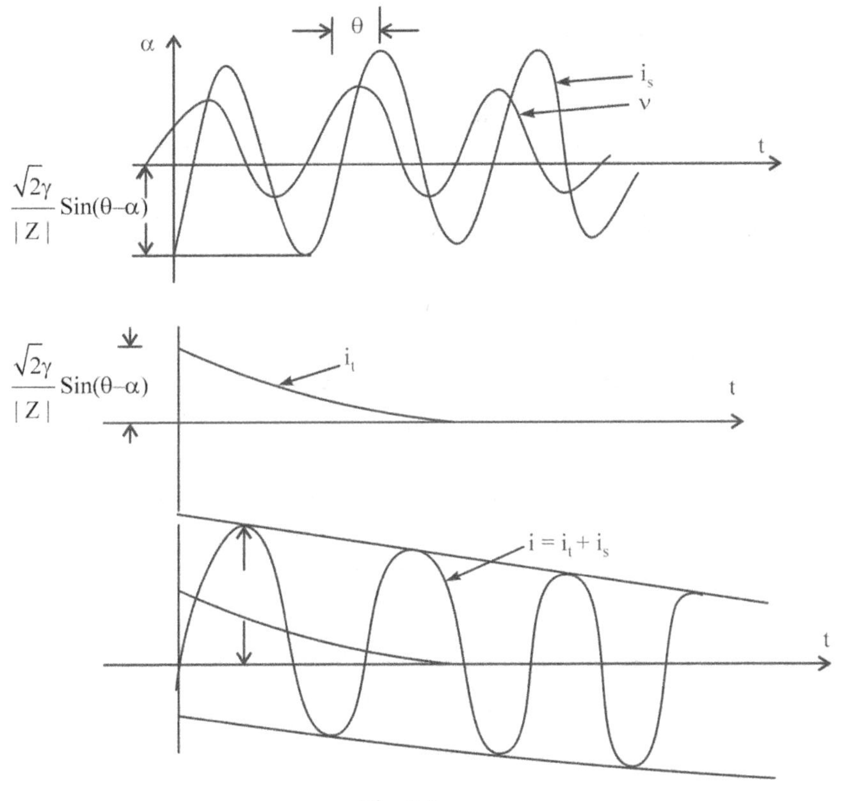

Fig.2.3

$$Ri(t) + L\,\frac{di(t)}{dt} = Vm\,\sin(\omega t + \propto) \qquad .....(2.1)$$

$$i(t) = \frac{V_m}{|z|}\,\sin(\omega t + \propto - \theta) + \frac{V_m}{|z|}\sin(\theta - \propto)e^{-\frac{R}{L}t} \qquad .....(2.2)$$

In Eq.(2.2), it said from the circuit theory that the short circuit current consist of two components and they are sinusoidal steady state component and unidirectional transient component. The steady state current is known as symmetrical short circuit current and the transient component is known as DC off-set current.

The short circuit current i(t) may be given by equation 2.3(a)

$$i(t) = \frac{V_m}{|Z|}\sin\left(\omega t + \propto - \theta\right) + \frac{V_m}{|Z|}\sin\left(\theta - \propto\right)e^{-\left(\frac{R}{L}\right)t} \qquad \text{.....2.3(a)}$$

In short circuit current i(t), the value refer to the first peak is called the maximum momentary short circuit current i_{mm}. The first peak value is obtained when $\sin\left(\omega t + \propto - \theta\right) = 1$. by equation 2.3(b)

$$\therefore \qquad i_{(t)} = \frac{V_m}{|Z|} + \frac{V_m}{|Z|}\sin\left(\theta - \propto\right)e^{\frac{R}{L}t} \qquad \text{.....2.3(b)}$$

If the decay of transient current in the interval between t = 0 and time at which first peak occurs is neglected, then Eq. 2.3(b) may be expressed as:

$$i_{mm} = \frac{V_m}{|Z|} + \frac{V_m}{|Z|}\sin\left(\theta - \propto\right) \qquad \text{.....(2.4)}$$

In transmission lines, the resistance is very low, compared to reactance and so θ is nearly 90°.

$$\therefore \qquad i_{mm} = \frac{V_m}{|Z|} + \frac{V_m}{|Z|}\sin\left(90° - \propto\right)$$

$$\Rightarrow \qquad i_{mm} = \frac{V_m}{|Z|} + \frac{V_m}{|Z|}\cos\propto \qquad \text{.....(2.5)}$$

We say that, i_{mm} has maximum, possible value when $\propto = 0$. This implies that the effect of short circuit will be severe if the fault happens as the voltage wave is passing from zero.

When $\propto = 0$,

$$i_{mm(max)} = \frac{V_m}{|Z|} + \frac{V_m}{|Z|} = 2\frac{V_m}{|Z|} \qquad \text{.....(2.6)}$$

The max value of symmetrical short circuit current is $\frac{Vm}{|Z|}$. We can observe that the maximum possible value of max momentary short circuit is two times the value of the maximum symmetrical short circuit current. If such a condition exists in a transmission line then this consequence is known as doubling effect.

A safer choice of momentary current rating of circuit breaker can be maximum possible value of greatest momentary short circuit current. The interrupting current rating of the circuit breaker (CB) can be obtained by multiplying the symmetrical short circuit current by a suitable factor. The multiplication by a constant is necessary to account for the DC offset current on the instance of interruption.

2.2 Transients owing to a 3–Phase Alternator Terminal Short Circuit

Consider a 3-phase alternator running on no–load. If a 3–phase fault takes place at the terminals of an alternator, then a heavy short circuit flows in the immature circuit.

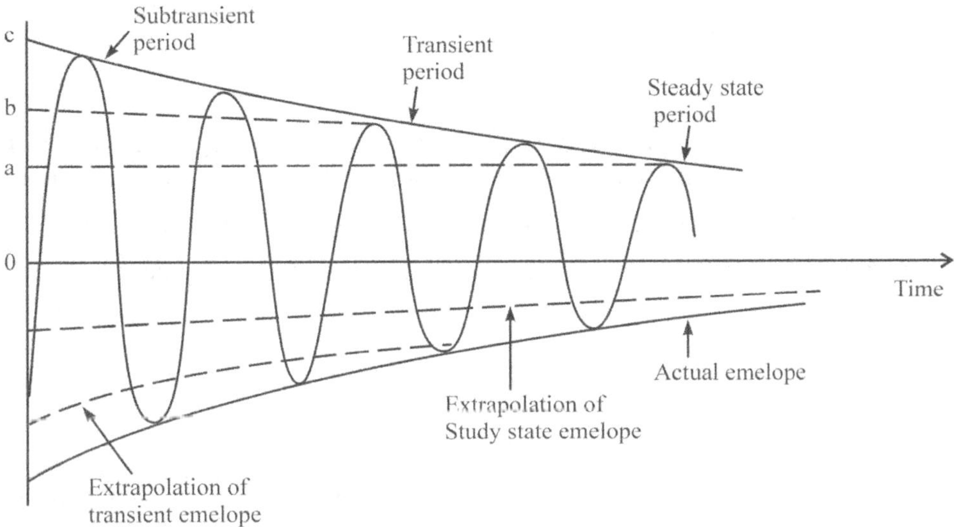

Fig.2.4

The symmetrical short circuit current may be classified into three sections known as:

(a) Sub transient.

(b) Transient

(c) Steady state.

Under steady state short circuit states demagnetizing flux is generated due to the armature reaction of a synchronous generator. This effect is represented as a reactance called armature reaction. Armature reactance is denoted by x_a. The sum of leakage reactance, (x_c) and Armature reactance (x_a) is called synchronous reactance, x_s.

For a salient pole machine, the synchronous reactance is called direct axis reactance and denoted by x_d.

At the instant of a short circuit, the DC-offset current appears in each of the three phases of the stator. This DC-offset current induces currents in the rotor field winding and damper winding by transformer action. The increase in the field current and damper winding current will set up flux in a direction to argument the main flux.

This effect may be signified by two reactances in parallel with x_a. Here x_f represents the flux created by the induced current in the field winding and x_{dw} represents the flux created by induced currents in the damper winding.

The joint consequences of all the three reactances are to reduce the total reactance of the machine and so the short circuit current is very high in this state which is called as sub-transient state. The total reactance under this condition is called sub – transient reactance and denoted by x_d'' given in equation (2.7)

$$x_d'' = x_L + \cfrac{1}{\cfrac{1}{x_a} + \cfrac{1}{x_f} + \cfrac{1}{x_{d\omega}}} \qquad .....(2.7)$$

The induced currents in the damper winding disappear after few cycle from the instant of fault, because the time constant of the damper winding is smaller than the field winding.

This effect is equivalent to open circuited x_{dw} and this state is called transient state.

The total reactance in the transient state is known as transient reactance and is indicated by x_d''.

$$x_d'' = x_1 + \cfrac{1}{\cfrac{1}{x_a} + \cfrac{1}{x_f}} \qquad .....(2.8)$$

The transient state will exist for few cycles and then steady state conditions are achieved. Because, the effect of the field winding current will also die-out in a short time depending on its time constant. This effect is equivalent to open circuited x_f and this state is referred to as steady state.

In steady state the total reactance is provided by a sum of x_1 and x_a. Thus Steady state reactance is given by x_d as shown in equation (2.9)

$$x_d = x_L + x_a \qquad ....(2.9)$$

From equations (2.7), (2.8) and (2.9) it is observed that the sub transient reactance of the machine is the smallest, and steady state reactance of the machine is highest among the reactances.

So, $\qquad\qquad x_d'' < x_d' < x_d$

Exercise

1. What do you mean by Power System Transient Fault?

2. Why does fault occurs in Power system?

3. Derive an expression to determine the different types of currents due to Transients due to Balanced Fault.

4. Explain the following different types of currents during short circuit:

   a) Sub transient   b) Transient   c) Steady state.

Chapter 3

Symmetrical Fault Analysis

3.1 Necessity for Short Circuit Calculation

Short circuit evaluation or fault calculations are to be done in power systems for determine the magnitudes of currents flowing for the duration of the power system at a variety of time intervals after a fault occurs. It additionally requires choosing the rating for fuses and protective gears and coordination of relay. It is helpful to build the cable joints consequently in power stations and in substations. Short circuit study is also ensured the MVA ratings of the existing circuit breakers when new generators are introduced into a system. Based on the short circuit analysis the systems grounding planning can be made correctly.

3.2 Symmetrical Fault Calculation

As discussed in the previous chapter, there are two types of faults in power system. In the present section we will deal with symmetrical fault calculation using per unit calculation. To solve the symmetrical fault calculation the following important sections are useful for analytical solution of any numerical problems.

3.3 Application of Thevenin Theorem

Any power network can be presented as a pair of terminals by a simple equivalent network consisting of a voltages source V_{TH} in series with an impedance Z_{TH}. The expression for fault current is given by

$$I_f = \frac{V_{TH}}{Z_{TH} + Z_f}$$

where Z_f is the fault impedance.

3.3.1 Symmetrical Short Circuit Analysis

Short Circuit Capacity SCC

In a Power network, short circuit capacity (SCC) of a bus may be expressed as the product of the magnitudes of the pre-fault voltage and post fault current. The SCC is also referred as the fault level.

$$|SCC| = |V^0| \times |I_F| \text{ VA} \qquad\qquad .....(3.1)$$

Where, V^0 = pre-fault voltage in Volts.

I_F = post fault currents in Amps

In case of a solid fault impedance $Z_f = 0$ and the fault current is given by

$$I_f = \frac{\left|V_{TH}\right|}{\left|Z_{TH}\right|}$$

$$\ldots\ldots(3.2)$$

where V_{TH} = Thevenin's voltage per phase in Volts

$\quad Z_{TH}$ = Thevenin's impedeance in Ohms.

Let us assume, $V^0 = V_{TH}$

$$\therefore \quad \left|SCC\right|_{1-\varphi} = \left|V_{TH}\right|\left|I_F\right| = \frac{\left|V_{TH}\right|^2}{Z_{TH}}$$

The Thevenin's impedance may be expressed in per unit by:

$$\left|Z_{TH}\right|_{p.u.} = \frac{\left|Z_{TH}\right|}{Z_{Base}} = \left|Z_{TH}\right| \times \frac{S_b}{V_b^2}$$

$$\ldots\ldots(3.3)$$

where S_b = Base Volt ampere in VA,

$\quad V_b$ = Base voltage in Volts

$\quad Z_{TH}$ = Thevenin's impedance in Ohms

Let us assume, base value $\left|V_{TH}\right| = V_b$

$$\therefore \quad \left|Z_{TH}\right|_{p.u.} = \left|Z_{TH}\right| \times \frac{S_b}{\left|V_{TH}\right|^2}$$

$$\frac{\left|V_{TH}\right|^2}{Z_{TH}} = \frac{S_b}{\left|Z_{TH}\right|_{p.u}}$$

$$\ldots\ldots(3.4)$$

From equations (3.3) and (3.4) we get,

$$\left|SCC\right|_{1-\varphi} = \frac{S_b}{\left|Z_{TH}\right|_{p.u}} \text{ VA/Phase}$$

$$\ldots\ldots(3.5)$$

In case of a three phase system, if $(KV_{TH/ph})$ is the phase voltage in KV then,

$$\left|SCC\right|_{1-\varphi} = \frac{\left|(KV_{TH})_{ph}\right|^2}{\left|Z_{TH}\right|} \text{ MVA /Phase}$$

Hence, for Total |SCC| for all three phases is given by, $|SCC|_{3-\varphi} = 3 \times |SCC|_{1-\varphi}$

$$|SCC|_{3-\varphi} = \frac{3\left|(KV_{TH})_{ph}\right|^2}{|Z_{TH}|} \, MVA$$

If $(KV_{TH})l$ is the line voltage in KV, then $(KV_{TH})l = \sqrt{3}\,(KV_{TH})_{ph}$

$$|SCC|_{3\varnothing} = \frac{\left|(KV_{TH})_l\right|^2}{|Z_{TH}|} \qquad\qquad .....(3.6)$$

Also, Z_{TH} is expressed in per unit value as,

$$|Z_{THp.u.}| = |Z_{TH}| \times \frac{S_{b.3\varnothing}}{\left|(KV_{TH})l\right|^2}$$

From which we get,

$$\frac{\left|(KV_{TH})_l\right|^2}{|Z_{TH}|} = \frac{S_{b.3\varnothing}}{Z_{THp.u.}} \qquad\qquad .....(3.7)$$

Using equation (3.6) and (3.7), we get

$$|SCC|_{3\varnothing} = \frac{S_{b.3\varnothing}}{Z_{THp.u.}} \, MVA$$

From the above equation, SCC can be found if the thevenin's per unit impedance is known.

The fault current (short circuit current) may be found as follows:

$$I_f = \frac{SCC_{3\varnothing} \times 10^6}{\sqrt{3} \times V_{TH} \times 10^3}$$

Example 1: Fig.(3.1) shows the single line diagram of 3 phase system. The per unit reactance of each alternator is based on its own capacity. Find the short circuit current that will flow in complete 3-phase short circuit at F.

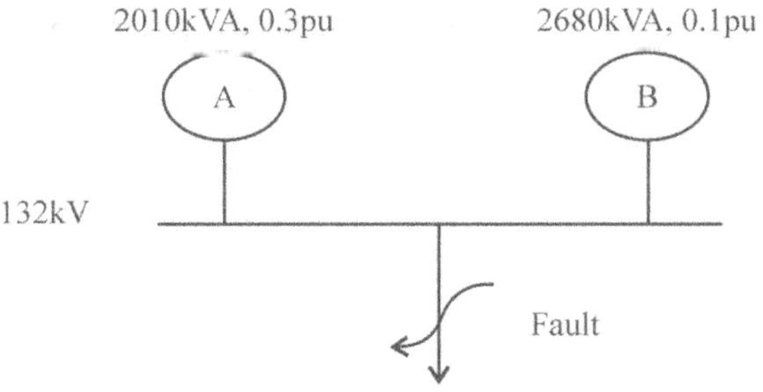

Fig.3.1

Solution:

Let us consider, the Base KVA = 2680 KVA

and Base KV = 132KV

Thus, p.u. reactance at Base KVA and Base KV level for generator A,

$$= 0.3 \times (2680/2010) \times (132/132)^2$$

$$= 0.4 \text{ p.u.}$$

Thus, p.u. reactance at Base KVA and Base KV level for generator B,

$$= 0.1 \times (2680/2680) \times (132/132)^2$$

$$= 0.1 \text{ p.u.}$$

The p.u. reactance diagram of the network is given in Fig.(3.2)

The Thevenin's equivalent (Z_{TH}) impedance, is $0.4 \times 0.1 /(0.4 + 0.1) = 0.04/0.5 = 0.08$ p.u.

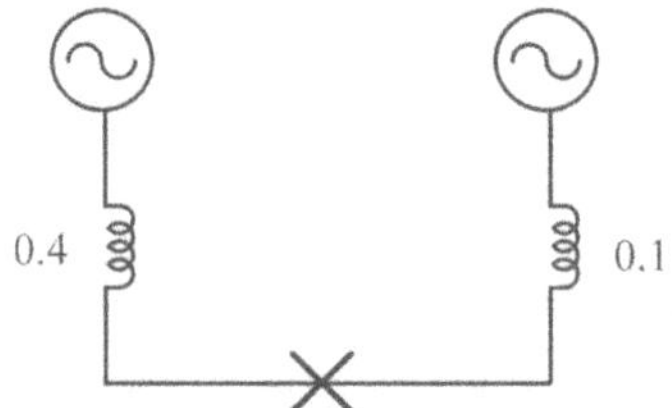

Fig.3.2

The short circuit capacity for the fault, $S_{cc} = \dfrac{S_b}{Z_{TH}}$ KVA $= \dfrac{2680}{0.08}$ KVA = 33500 KVA

The fault current at the fault point, $I_f = \dfrac{SCC_{cc}}{\sqrt{3}V_{LL}} = \dfrac{33500}{\sqrt{3}\times132} = 146.52$ Amp

Example 2: The estimated short-circuit MVA at the bus-bars of a generating station A is 1880 MVA and of another station B is 1440 MVA. The generated voltage at each station is 39.6 KV. If these stations are interconnected through a line having a reactance of 1.2 Ω and negligible resistance, calculate the possible short-circuit MVA at both stations.

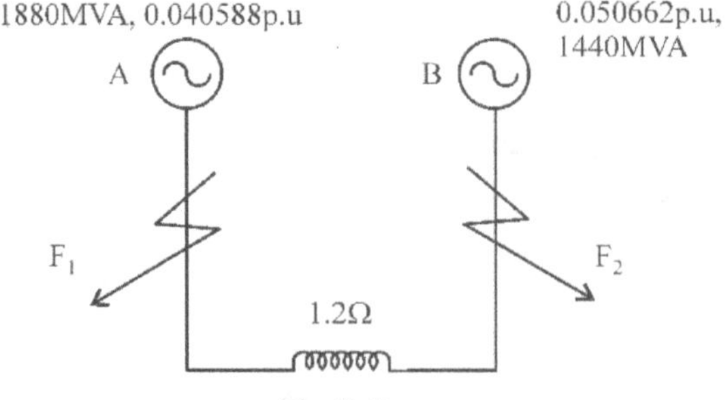

Fig.3.3

Now, given that

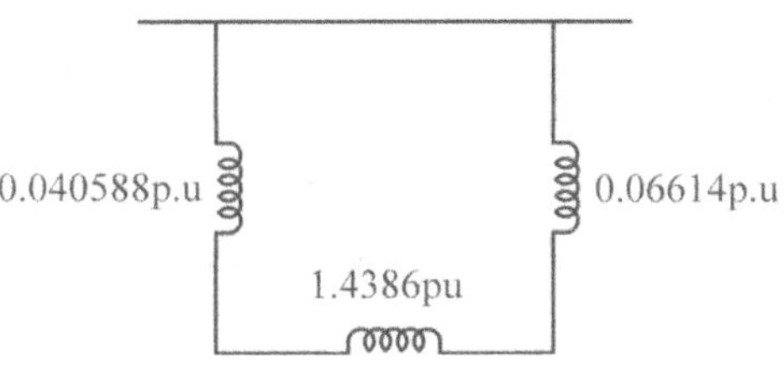

Fig.3.4

Let us consider, the Base KVA (Sb_2) = 1880MVA

And Base KV(KV_2) = 39.6 KV

Thus, p.u. reactance at Base MVA and Base KV level for generator 1,

$$= Z_{pu1} = Z_{pu} \times \frac{Sb_2}{Sb_1} \times \left(\frac{KV_1}{KV_2} \right)^2$$

$$= 0.0405 \times \frac{1880}{1880} \times \left(\frac{39.6}{39.6} \right)^2$$

$$= 0.040588pu$$

Thus, p.u. reactance at Base KVA and Base KV level for generator 2,

$$= Z_{pu2} = Z_{pu} \times \frac{Sb_2}{Sb_1} \times \left(\frac{KV_1}{KV_2} \right)^2$$

$$= 0.050662 \times \frac{1880}{1440} \times \left(\frac{39.6}{39.6} \right)^2$$

$$= 0.06614pu$$

Thus, p.u. reactance at Base KVA and Base KV level for reactance, $Z_{act} = 1.2 \ \Omega$

$$Z_{pu}. \text{ reactor} = Z_{act} \times \frac{Sb_2}{(Vb)^2}$$

$$= 1.2 \times \frac{1880}{(39.6)^2}$$

$$= 1.2 \times \frac{1880}{1568.16}$$

$$= 1.4386pu$$

For Case 1

The p.u reactance diagram of the network is given in Fig. (3.5)

The Thevenin's equivalent (Z_{Th}) impedance,

$$Z_{TH} = \frac{0.040588 \times (1.4386 + 0.06614)}{0.040588 + 1.4386 + 00.06614} = 0.0395 \text{ pu}$$

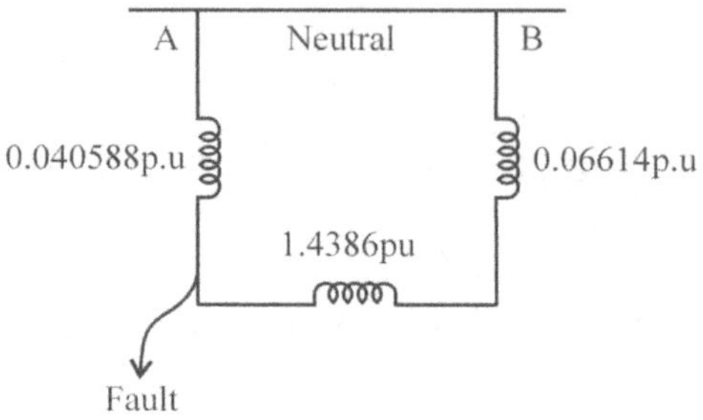

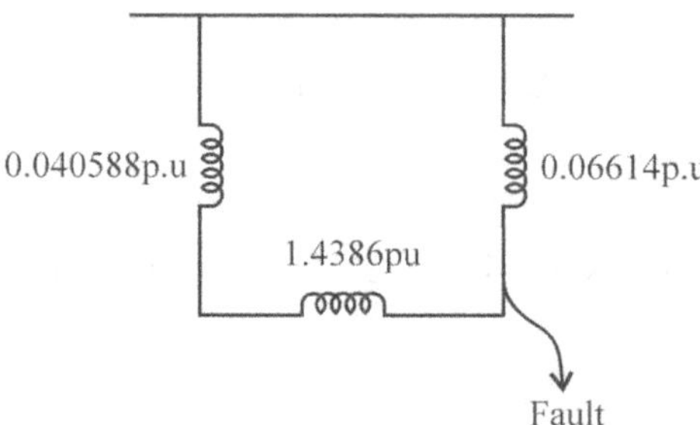

Fig.3.5

$$\text{Short Circuit Capacity (SCC) 3 phase} = \frac{Sb_3 \text{ phase}}{Z_{TH}} \text{ MVA}$$

$$= \frac{1880}{0.03958} \text{ MVA}$$

$$= 47594 \text{ MVA}$$

For Case 2

Fig.3.6

The p.u. reactance diagram of the network is given in Fig.3.6.

The Thevenin's equivalent (Z_{Th}) impedance,

$$Z_{TH} = \frac{00.06614 \times (0.040588 + 1.4386)}{(0.040588 + 1.4386 + 00.06614)} = 0.0633 \text{pu}$$

$$\text{SCC } 3-\varphi = \frac{S_b \, 3-\Phi}{Z\text{pu}} \text{ MVA}$$

$$= \frac{1880}{0.0633} \text{ MVA}$$

$$= 29699.84 \text{ MVA}$$

Example 3: A 3-phase transmission line operating at 33KV and having a resistance of 0.68Ω and reactance of 2.72Ω is connected to the generating station bus-bars through 5MVA step-up transformer having a reactance of 0.034p.u. The bus-bars are supplied by a 10MVA alternator having 0.068 p.u. reactance. A Load of 250 HP motor is running at a p.f of 0.866.Calculate the short-circuit kVA fed to symmetrical fault between phases if it occurs (i) at the load end of transmission line (ii) at the high voltage terminals of the transformer.

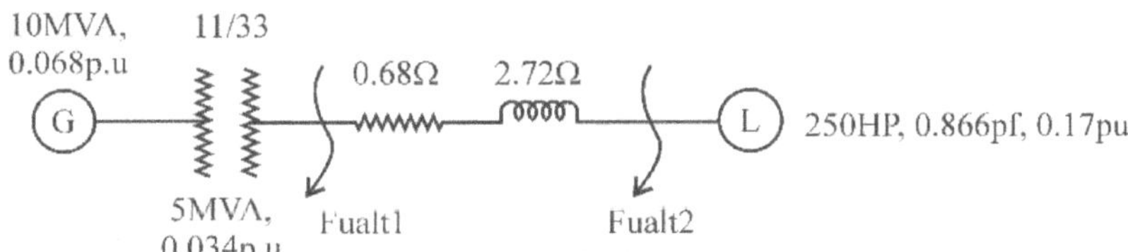

Fig.3.7

Solution:

Base MVA – 100 MVA

Base KVA – 11 KVA

Generator:

$$Zp.u.(new) = Zp.u.given \times \left(\frac{\text{Base KV given}}{\text{Base KV new}}\right)^2 \times \left(\frac{\text{Base MVA new}}{\text{Base MVA given}}\right)$$

$$Z\, p.u.(new) = 0.068 \times (100/10) \times (11/11)^2$$

$$= 0.68 \text{ p.u.}$$

Transformer:

$$Z_{p.u.(new)} = Z_{p.u.\,given} \times \left(\frac{\text{Base KV given}}{\text{Base KV new}}\right)^2 \times \left(\frac{\text{Base MVA new}}{\text{Base MVA given}}\right)$$

$$Z\, p.u.\ (new) = 0.034 \times (100/5) \times (11/11)^2$$
$$= 0.68 \text{ p.u.}$$

Transmission line:

$$Z\, p.u. = Z_{act} \times \text{Base MVA}/(\text{Base KVA})^2$$

$$= 0.68 + j2.72 \times 100/(33)^2$$

$$= 0.062 + j0.247$$

For Motor: 250 Hp, 0.866p.f., 0.17pu $= \dfrac{250 \times 0.746}{0.866}$ MVA $= 215.4$ MVA

Z_{M}p.u. $= Z_{(act)} \times$ Base MVA/Given Base MVA

$\qquad = 0.17 \times 100/215.4 = 0.0789$

Impedance Diagram

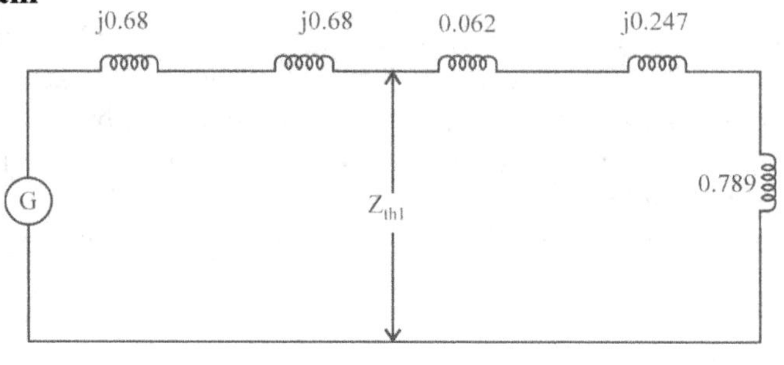

Fig.3.8

(i)   For Fault 1:

$$Z_{TH} = \frac{\left(j0.68 + j0.68\right) \times \left(j0.62 + j0.247 + j0.789\right)}{\left(j0.68 + j0.68 + j0.62 + j0.247 + j0.789\right)}$$

$Z_{TH} = (j0.60 + j0.68)$

$\qquad = j1.36$ ohm

SCC $(3\text{-}\varphi) = S_b(3\text{-}\varphi)/Z_{th}$p.u.

$\qquad\qquad = (100/j1.36)$

$\qquad\qquad = j73.52$

Fault current (IF) $=$ SCC $(3-\varphi) / \sqrt{3} \times V_{lb} \times 10^3$

$\qquad\qquad = (j73.52 \times 10^6)/(\sqrt{3} \times 33 \times 10^3)$

$\qquad\qquad = 1286.266$ A

(ii) For Fault 2:

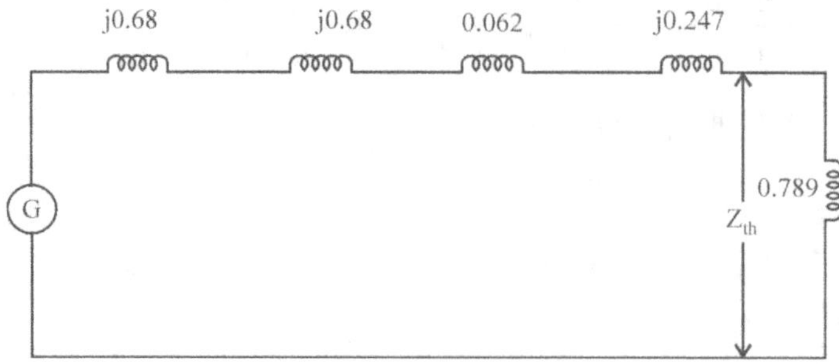

Fig.3.9

$$Z_{TH} = \frac{(j0.68 + j0.68) + (j0.62 + j0.247)(j0.789)}{(j0.60 + j0.68 + j0.62 + j0.247 + j0.789)} = \frac{(2.227) \times (0.789)}{3.016} = 0.5803 \text{ p.u.}$$

$$= \frac{100}{0.5803} = 172.32 \text{ MVA}$$

$$= 3014.988 \text{ Amp}$$

$$I_F = \frac{100 \times 10^6}{\sqrt{3} \times 33 \times 10^3} = 0.5803 \text{ MVA}$$

Example 4: The section bus-bars A and B are linked by a bus-bar reactor rated at 4000 KVA with 0.1pu reactance. On bus-bar A, there are two generators each of 15,800 KVA with 0.158 pu reactance and on B two generators each of 7000 KVA with 0.079pu reactance. Find the steady MVA fed into a dead short circuit between all phases on B with bus-bar reactor in the circuit.

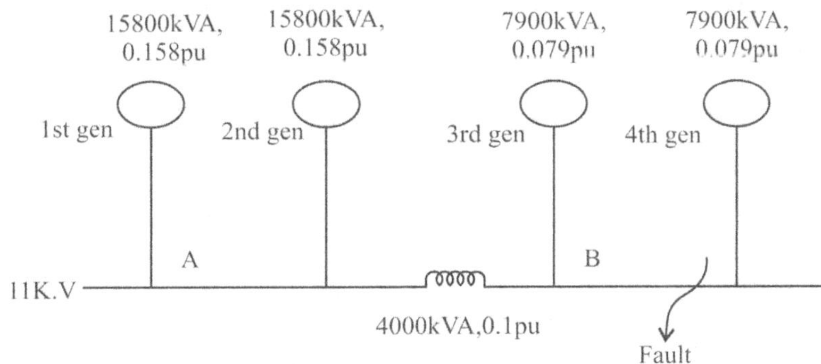

Fig.3.10

Solution:

Consider Base KVA to be 7900 and Base KV to be 11 KV i.e.:

$Sb_1 = 7900$ KVA

$Vb_1 = 11$KV

Formulae Used:

$$Z_{pu.2} = Z_{pu.1} \times (Sb_2/Sb_1) \times (Vb_1/Vb_2)^2$$

$$S_{scc,3phase} = S_{b,3phase}/Z_T \text{ pu.}$$

Calculation For Generator 4:

$$Z_{pu.4\,G} = 0.079 \times (7900/7900) \times (11/11)^2$$

$$= 0.079 \text{ p.u.}$$

Calculation for Generator 3:

$$Z_{pu.3G} = 0.079 \times (7900/7900) \times (11/11)^2$$
$$= 0.079 \text{ p.u.}$$

Calculation For Generator 2:

$$Z_{pu.2\,G} = 0.158 \times (7900/15800) \times (11/11)^2$$
$$= 0.079 \text{ p.u.}$$

Calculation For Generator 1:

$$Z_{pu.1G} = 0.158 \times (7900/15800) \times (11/11)^2$$
$$= 0.079 \text{ p.u.}$$

Calculation for Transmission Line:

$$Z_{puTM} = 0.1 \times (7900/4000) \times (11/11)^2$$
$$= 0.1975 \text{ p.u.}$$

Impedance Diagram

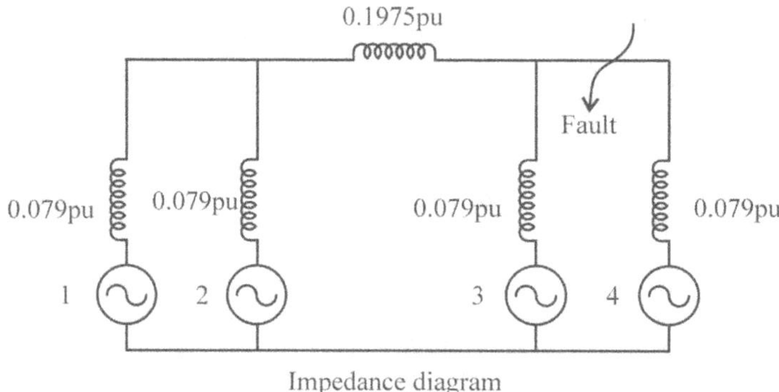

Impedance diagram

Fig.3.11

$$Z_{T,pu} = \{(0.079/0.079) + 0.1975] (0.079\ 0.079)$$
$$= 0.1975 \text{ p.u}$$

$$S_{SCC,\,3phase} = S_{b,3phase} / Z_T \text{ pu.}$$
$$= 7900 / 0.03385$$
$$= 23338.25 \text{ KVA}$$
$$= 23.338 \text{ MVA}$$

So, The steady MVA fed into the dead short circuit will be 36.363 MVA.

1. **Selection of circuit Breakers:** The circuit breaker is an important protective device used in medium and high voltage (high level power) of power system in transmission lines,

substations, generating stations and for heavy loads in industries to isolate the faulty element of the power system during the fault.

Both in fault condition and load conditions, heavy current will interrupted by its contactor for its opening condition. There will be large voltage with sparking appeared across the contactor because of inductive reactive type of major components in power system.

The circuit breaker for a particular load is selected based on the following ratings:

1. Usual operating power level identified as rated interrupting current or rated interrupting KVA.
2. The fault level identified by rated short circuit interrupting current or rated short circuit interrupting MVA.
3. Momentary current rating.
4. Usual operating voltage.
5. Speed of circuit breaker.

The speed of circuit breaker is defined by the time between the incidences of the fault to the extinction of the arc (when the contact opens). It is in general identified in cycles of power frequency. The benchmark speed of circuit breakers are 8, 5, 3 or 11/2 cycles.

2. **Reactor:** Reactor is a coil with high inductive reactance as contrasted to its resistance. It may be employed to limit the short circuit current all through fault situations. To execute short circuit limit important property for a reactor is magnetic saturation at huge current does not decrease the coil reactance. Such type of iron cored inductor is very expensive and weighty. So air-cored coils with constant inductance are usually employed for current limiting reactors.

Air cored reactors are usually of two categories:-

(1) oil immersed type

(2) dry type.

Oil immersed reactors may be cooled in any way exercised for cooling the power transformer whereas the dry type are normally cooled by natural exposure to air and are sometimes planned with forced-air and heat exchanger auxiliaries. Reactors are normally constructed for single phase units.

With the raise in interconnection of power system, the fault level is growing. It is, therefore, essential to enhance the reactance by bring in reactors at intentional points in the system.

3.4 Generator Reactors

The reactance of modern alternator may be as high as 2.0 p.u. which means even a dead short circuit at the terminals of the alternator will result in a current less than full load current and no external reactor is required for limiting the short circuit current of such a machine.

3.5 Feeder Reactor

The per unit value of reactance of a feeder based on its ratings may be small but when compared with the rating of the whole system, its value is quite large and hence a small reactor will be useful in reducing the short circuit current should fault occur close to the generating station. In case this feeder reactor is not there, a fault in such a location would bring the bus bar voltage almost down to zero value and there is a possibility of various generators falling out of step.

3.6 Bus Bar Reactor

There are three methods of interconnecting the bus bar through the reactors as illustrated in Fig.(3.12).

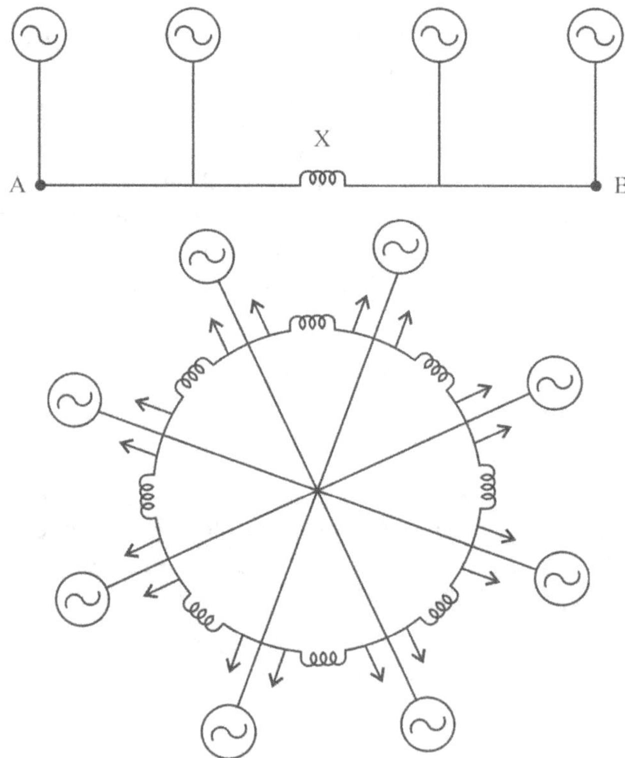

Fig. 3.12

The simple method is suitable for plants of moderate output whereas for large sized plants either the star or ring system of connection is used. It is to be noted that any transfer of power from any section A to section B of the generators, a difference in potential between the bus sections is developed. If the power to be transferred is watt less the difference in voltage between the bus sections will be transferred.

Exercise

Short Questions

1. The power system is subjected to a fault which makes the zero-sequence component of current equal to zero. The nature of fault is :
   a) Double line to ground fault
   b) Double line fault
   b) Line to ground fault
   d) Three phase to ground fault

2. Possible faults may occur on a transmission line are
   a) 3-phase fault
   b) L-L-G fault
   c) L-L fault
   d) L-G fault

3. The decreasing order of severity of the faults from the stability point of view is:
   a) 1-2-3-4
   b) 1-4-3-2
   c) 1-3-2-4
   d) 1-3-4-2

4. _______ is a series type unbalanced fault that occurs in a power system.
   a) Line – to – line fault
   b) Double line – to – ground fault
   c) Single line – to – ground fault
   d) Open conductor fault

Board Questions

1. Explain the significance of using circuit breaker in power system. Describe different types of CB.

2. What is the importance of short-circuit calculations ?

3. What is the importance of base kVA in short-circuit calculations ?

4. Why do we choose a base kVA in short-circuit calculations ?

5. Why do we decide the <u>rating</u> of a circuit breaker on the basis of symmetrical short-circuit currents?

6. Explain the importance of reactor in power system. Explain the different types of reactors.

Answers

1. a,   2. a,   3. d

Chapter 4

The Analytical Concept for
Symmetrical Components

4.1 Introduction

The analysis of unsymmetrical polyphaser network by employing symmetrical components was introduced by Dr. C. L. Fortesque. He proved that an unbalance system of 'n' related vectors may be determined into n system of balanced vectors known as 'symmetrical components' of original vectors. The n vectors of each set of components are identical length and the phase angles among adjacent vectors of the set are the same.

4.1.1 Concept of 'a' Operator

For an unbalanced power network, the current and voltage of each phases are different and there not 120^0 phase apart to each other. Thus, each phase can be expressed in terms of symmetrical components. In this idea a positive, negative and zero sequence sets are uniformly spaced apart by 120° each. Consequently an operator is needed that will cause 120° rotation. Thus a operator is utilised that assists in the revolving of vectors by 120° in anticlockwise path when multiplied to a vector.

If a vector V in OP position is multiplied with a operator then,

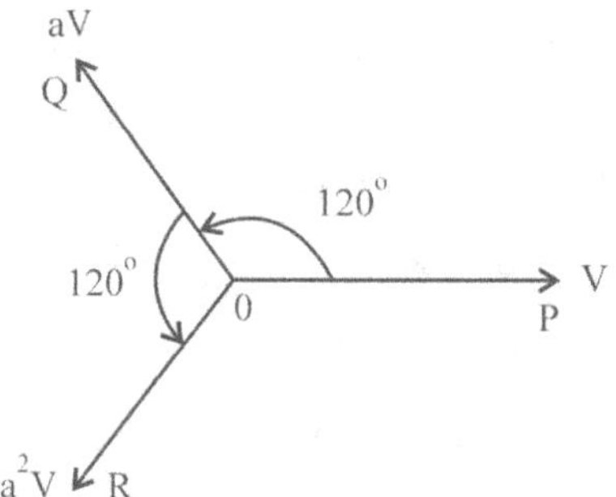

Fig.4.1

$$aV = V < 120°$$
$$= V(-0.5 + j\,0.866)$$
$$a = -0.5 + j\,0.866 \qquad\qquad .....(4.1)$$

When a vector in position OQ is multiplied by a operator then,

$$a^2 V = V < 240°$$

$$= V(-0.5 - j\,0.866)$$

$$a^2 = -0.5 - j\,0.866 \qquad .....(4.2)$$

Similarly,

$$a^3 = 1$$

2. Properties of 'a' operator.

   From the above equations we know,

$$a = -0.5 + j\,0.866 \qquad .....(4.3)$$

$$a^2 = -0.5 - j\,0.866 \qquad .....(4.4)$$

1. summing these two equations we get,

$$1 + a + a^2 = 0$$

2. subtracting these two equations we get,

$$a - a^2 = j\,\sqrt{3}$$

4.2 Sequence Component to Unbalance Phase Conversion

In a three phase system, the three unbalanced vectors (I_{R1}, I_{R2}, I_{R0}) may be determined into three balanced system of vectors. The vectors of the balanced system are known as symmetrical components of the original system. The symmetrical components of three phase system are:

1. Positive sequence components.
2. Negative sequence components.
3. Zero sequence components.

The positive sequence components having three identical vectors in magnitude, displaced from each other by 120° in phase with the same phase sequence as the original vectors. The negative sequence components also having three identical vectors equal in magnitude, displaced from each other by 120° in phase with the phase sequence opposite to that of the original vectors. The zero sequence components having three vectors equal in magnitude and with zero phase displacement from each other.

In 3-phase system, we can express the phase current in terms of symmetrical component through the following equations:

$$I_R = I_{R1} + I_{R2} + I_{R0} \qquad .....(4.5)$$

$$I_Y = I_{Y1} + I_{Y2} + I_{Y3} \qquad .....(4.6)$$

$$I_B = I_{B1} + I_{B2} + I_{B3} \qquad .....(4.7)$$

There are three sets of independent components in power system. Positive, negative and zero sequence. Positive sequence components are always present in the system. A second set of balanced phase's equivalent magnitude but displaces 120° apart is known as negative sequence. The third set of phases identical in size and in same phase with each other is identified as zero sequence. The sequence components can be used to determine any unbalanced current or voltage. We can observe the symmetrical components being represented in the above figure.

Let us now express the symmetrical parts of R phase by means of phase currents. These can be done by denoting the symmetrical components of Y and B in the form of R with the help of 'a' operator.

Now,

$$I_R = I_{R1} + I_{r2} + I_{r0} \qquad\qquad .....(4.8)$$

$$I_Y = I_{Y1} + I_{Y2} + I_{Y0}$$

$$= a^2\, I_{R1} + a\, I_{R2} + I_{R0} \qquad\qquad .....(4.9)$$

$$I_B = I_{B1} + I_{B2} + I_{B0}$$

$$= a\, I_{R1} + a^2\, I_{R2} + I_{R0} \qquad\qquad .....(4.10)$$

4.3   Unbalance Phase to Sequence Component Conversion

(i)   Zero sequence current.

Now adding the equations (4.8), (4.9) and (4.10), we get,

$$I_R + I_Y + I_B = I_{R1}(1 + a^2 + a) + I_{R2}(1 + a^2 + a) + 3I_{R0}$$

$$\therefore \qquad I_{R0} = \frac{1}{3}(I_R + I_Y + I_B)$$

Let us find the reference phase; therefore, subscript R is usually omitted.

$$\therefore \qquad I_0 = \frac{1}{3}(I_R + I_Y + I_B)$$

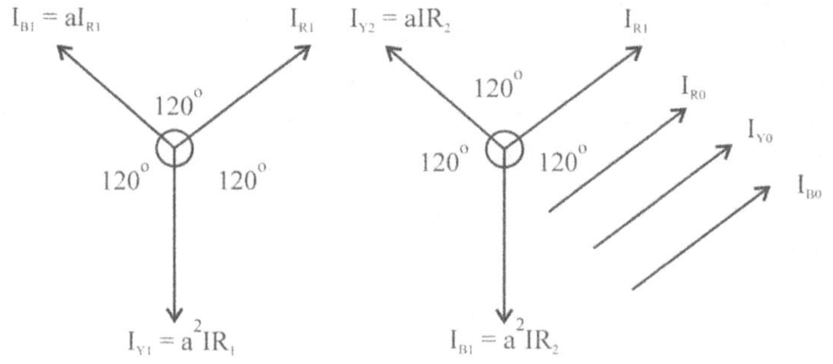

Fig.4.2

(ii) Positive sequence current:

Now multiply expression eq.4.9 by 'a' and exp. eq.4.10 by 'a2' and then adding these expression to expression eq.4.8, we get,

$$I_R + a\,I_Y + a^2\,I_B = I_{R1}\,(1 + a^3 + a^3) + I_{R2}\,(1 + a^2 + a^4) + I_{R0}\,(1 + a + a^2)$$

$$= 3\,I_{R1}$$

$$\therefore \qquad I_{R1} = \frac{1}{3}\,(I_R + a I_Y + a^2 I_B)$$

Omitting the subscript R, we have,

$$I_1 = \frac{1}{3}\,(I_R + a\,I_Y + a^2 I_B)$$

(iii) Negative sequence current:

Now, multiply expression eq.4.9 by 'a²' and expression eq.4.10 by 'a' and then adding this expression to eq.4.8, we get,

$$I_R + a^2\,I_Y + a\,I_B = I_{R1}\,(1 + a^4 + a^2) + I_{R2}\,(1 + a^3 + a^3) + I_{R0}\,(1 + a^2 + a)$$

$$= 3\,I_{R2}$$

$$\therefore \qquad I_{R2} = \frac{1}{3}\,(I_R + a^2 I_Y + a\,I_B)$$

Omitting the subscript R, we have,

$$\therefore \qquad I_2 = \frac{1}{3}\,(I_R + a^2 I_Y + a\,I_B)$$

The above relation between sequence and phase components also applies two voltages. Similarly for voltage components the relation will be as follows:

$$E_0 = \frac{1}{3}\,(E_R + E_Y + E_B)$$

$$E_1 = \frac{1}{3}\,(E_R + a\,E_Y + a^2\,E_B)$$

$$E_2 = \frac{1}{3}\,(E_R + a^2\,E_Y + a\,E_B)$$

Example 1: A star connected balanced load takes 108 A from a balanced 3-phase, 4-wire supply. Find the symmetrical components of the line currents at the following conditions:

(i)   Before the fuses are removed in the Y and B phases

(ii) After the fuses are removed in the Y and B phases

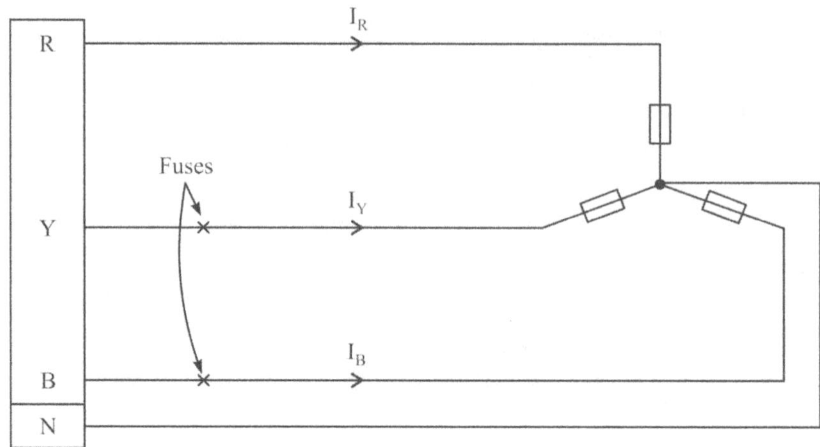

Fig.4.3

Solution: (i) Before the removal of fuses

Before removing the Y and B lines, the star connected 3-phase system is balanced and in its line is 108A.

$$I_R = 108 < 0 \text{ A} \qquad I_Y = 108 < 240 \text{ A} \qquad I_B = 108 < 120\text{A}$$

Since we already know that the system is balanced, it will only have positive sequence currents so negative sequence and zero sequence currents will be zero in the lines.

So,

$$I_{R0} = I_{Y0} = I_{B0}$$

$$= 1/3[I_R + I_y + I_B]$$

$$= \frac{1}{3}[108 < 0 + 108 < 240 + 108 < 120] = 0 \text{ A}$$

$$I_{R2} = 1/3[I_R + a^2 I_Y + aI_B]$$

$$= \frac{1}{3}[108 < 0 + 1 < -120 \times 108 < 240 + 1 < 120 \times 108 < 120]$$

$$= \frac{1}{3}[108 < 0 + 108 < 120 + 108 < 240]$$

$$= 0\text{A}$$

$$I_{Y2} = aI_{R2} = 1 < 120 \times 0 = 0\text{A}$$

$$I_{B2} = a^2 I_{R2} = 1 < 120 \times 0 = 0\text{A}$$

Hence, negative sequence and zero sequence components in the three lines are zero. So we can easily determine the value of the following sequence components.

$$I_{R1} = I_R = 108 < 0\text{A}$$

$I_{Y1} = I_Y = 108 < 240A$

$I_{B1} = I_B = 108 < 120A$

(i)   After removing the fuse:

When the fuse is removed in Y and B phases, the system is unbalanced and current in Y and B lines is zero.

$I_R = 108 < 0A$ $\qquad\qquad$ $I_B = I_Y = 0A$

So the sequence current in 3-phase line can be found out as such:

Zero sequence components

$$I_{R0} = I_{YO} = I_{B0}$$

$$= 1/3[I_R + I_B + I_Y]$$

$$= \frac{1}{3}[108 < 0 + 0 = 0] = 36 < 0A$$

Positive sequence components

$$I_{R1} = 1/3[I_R + aI_Y + a^2I_B]$$

$$= \frac{1}{3}[108 < 0 + 0 + 0] = 36 < 0A$$

$$I_{Y1} = a^2I_{R1} = 1 < 240 \times 36 < 0 = 36 < 240A$$

$$I_{B1} = aI_{R1} = 1 < 120 \times 36 < 0 = 36 < 120 \text{ A}$$

Negative sequence components

$$I_{R2} = 1/3[I_R + a^2I_Y + aI_B]$$

$$= \frac{1}{3}[108 < 0 + 0 + 0] = 36 < 0A$$

$$I_{Y2} = aI_{R2} = 1 < 120 \times 36 < 0 = 36 < 120A$$

$$I_{B2} = a^2I_{R2} = 1 < 240 \times 36 < 0 = 36 < 240A$$

Example 2: $3\emptyset$, 4-wire system is operating under unbalance condition. During the unbalancing it was observed that R phase positive sequence and the negative sequence component were $240 < 0A$ and $120 < 72$. A respectively was also observed that $360 < 360$ A current flowing back to the supply through neutral conductor. Determine the current in the three lines.

Solution: Total current flowing in the neutral wire $= 360 < 360$

$$I_{RO} = \frac{1}{3} \times \text{Current in neutral line}$$

$$I_{RO} = \frac{1}{3} \times 360 < 360$$

$$= 120 < 360$$

And so as given, $I_{R1} = 240 < 0A$

$\quad I_{R2} = 120 < 72A$

Current in R-Line

$\quad I_R = [I_{RO} + I_{R1} + I_{R2}]$

$\quad\quad = 120 < 360 + 240 < 0 + 120 < 72$

$\quad\quad = 120(\cos0 + \sin360) + 240(\cos0 + \sin0) + 120(\cos72 + \sin72)$

$\quad\quad = 120 + 240 + 37.082 + j114.12$

$\quad\quad = 413.15 < 16A$

Current in Y- Line

$\quad I_Y = I_{RO} + a^2 I_{R1} + a\ I_{R2}$

$\quad\quad = 120 < 360 + 1 < 240 \times 240 < 0 + 1 < 120 \times 120 < 72$

$\quad\quad = 120 + 240 < 240 + 120 < 192$

$\quad\quad = 120 + (-120 - j207.84) + (-117.377 - j24.949)$

$\quad\quad = -117.377 - j232.789$

$\quad\quad = 260.706 < 63.305\ A$

Current in B-Line

$\quad I_B = I_{R0} + aI_{R1} + a^2 I_{R2}$

$\quad\quad = 120 < 360 + 1 < 120 \times 240 < 0 + 1 < 240 \times 120 < 72$

$\quad\quad = 120 + 240 < 120 + 120 < 312$

$\quad\quad = 120 + (-120 + j207.84) + (80.29 - j89.17)$

$\quad\quad = 80.29 + j118.67 = 143.27 < 55A$

Example 3: Fig.4.4 shows the current is flowing to the Δ-connected load from a 3 phase supply. B phase terminal is open; determine the symmetrical components of the line currents.

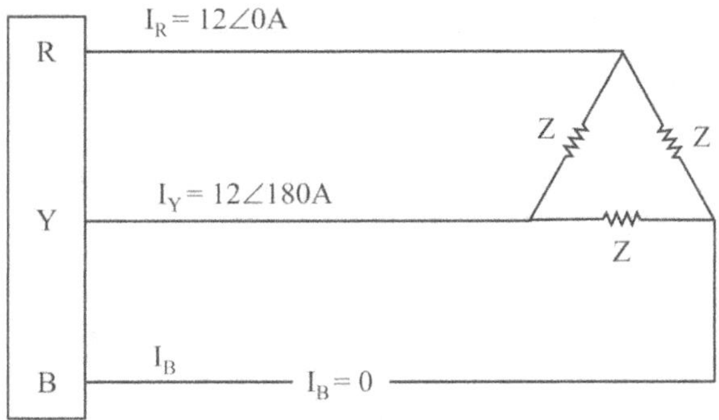

Fig.4.4

Solution: As given that the one of the conductor of the 3-phase line is open and current in R-line is 12A.

$$I_R = 12 < 0A \qquad I_Y = 12 < 180A \qquad I_B = 0A$$

So the sequence component of the R-line:

$$I_{R0} = 1/3[I_R + I_Y + I_B]$$

$$= \frac{1}{3}[12 < 0 + 12 < 180 + 0] = 0A$$

$$I_{R1} = 1/3[I_R + aI_Y + a^2I_B]$$

$$= \frac{1}{3}[12 < 0 + 1 < 120 \times 12 < 180 + 0]$$

$$= \frac{1}{3}[12 + 12 < 300 + 0]$$

$$= 6 - j3.4641$$

$$= 6.928 < -29.99A$$

$$I_{R2} = 1/3[I_R + a^2I_Y + aI_B]$$

$$= \frac{1}{3}[12 < 0 + 1 < 240 \times 12 < 180 + 0] = 6 + j3.464$$

$$= 6.928 < 29.99A$$

Y-Line $\qquad\qquad I_{Y0} = I_{R0} = 0A$

$$I_{Y1} = a^2I_{R1} = 1 < 240 \times 6.928 < -29.99$$

$$= 6.928 < -150A$$

$$I_{Y2} = aI_{R2} = 1 < 120 \times 6.928 < 29.99$$

$$= 6.928 < 150A$$

B-Line $\qquad\qquad I_{B0} = I_{R0} = 0A$

$I_{B1} = aI_{R1} = 1 < 120 \times 6.928 < -29.99$

$\qquad = 6.928 < 90A$

$I_{B2} = a^2 I_{R2} = 1 < 240 \times 6.928 < 29.99$

$\qquad = 6.928 < -90A$

$\qquad = \dfrac{1}{3}(0) = 0$

Example 4: A Delta connected load of resistors of 6Ω, 12Ω and 24Ω are supplied power by a balanced 120 volts supply. Determine the sequence components currents in the resistors and in supply lines?

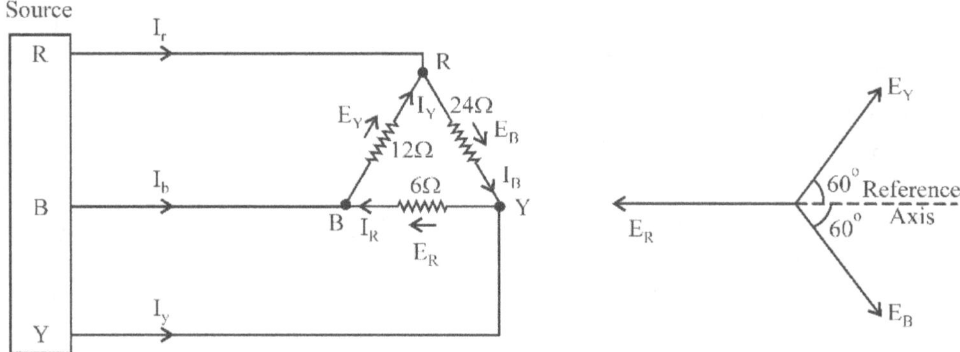

Fig.4.5

Solution:

Voltage and current across the 6ohm, 12ohm and 24ohm resistors are E_R, E_Y, E_B, I_R, I_Y and I_B respectively.

From the vector diagram, the voltages can be represented as such:

$E_R = -120 < 0V \qquad\qquad E_Y = 120 < 60V \qquad\qquad E_B = 120 < -60V$

And the currents are: $\quad I_R = E_R/6 = \dfrac{(-120 < 0)}{6} = -20 < 0A$

$$I_Y = E_Y/12 = \dfrac{-120 < 60}{12} = -10 < 60A$$

and $\qquad\qquad I_B = E_B/24 = \dfrac{120 < -60}{24} = 5 < -60A$

Sequence current in resistors:

Zero sequence current of I_R is

$I_{RO} = 1/3[I_R + I_Y + I_B]$

$$= \frac{1}{3}[-20{<}0 + 10 < 60 + 5 < -60]$$

$$= \frac{1}{3}[(-20 + j0) + (5 + j8.66) + (2.5 - j4.35)]$$

$$= -4.17 + j1.44 = 4.41 < -160.9A$$

Positive sequence component of I_R

$I_{R1} = 1/3[I_R + aI_Y + a^2I_B]$

$$= \frac{1}{3}[-20 < 0 + 1 < 120 \times 10 < 60 + 1 < 240 \times 5 < -60]$$

$$= 11.66 + j0 = 11.66 < 180A$$

Negative sequence component of I_R

$I_{R2} = 1/3[I_R + a^2I_Y + aI_B]$

$$= \frac{1}{3}[-20 < 0 + 1 < 240 \times 10 < 60 + 1 < 120 \times 5 < -60]$$

$$= 4.17 - j1.44 = 4.4 < -160.9A$$

By the help of I_R sequence component we can also find out the sequence component of I_Y and I_B.

So Y-Line components:

$I_{Y0} = I_{RO} = 4.41 < 160.9A$

$I_{Y1} = a^2I_{R1} = 1 < 240 \times 11.66 < 180 = 11.66 < 60A$

$I_{Y2} = aI_{R2} = 1 < 120 \times 4.4 < -160.9 = 4.41 < 110.9A$

B-Line components:

$I_{BO} = I_{R0} = 4.41 < 160.9A$

$I_{B1} = I_{R2} = aI_{R1} = 1 < 120 \times 11.66 < 180 = 11.66 < 300A$

$I_{B2} = a^2I_{R2} = 1 < 240 \times 4.4 < -160.9 = 4.4 < 79.1A$

So the current in R-Line $I_r = I_B - I_Y = 5 < -60 -10 < 60$

$$= 2.5 - j4.33 - 5 - j8.66$$

$$= -2.5 - j12.996 = 13.22 < -100.9A$$

Y-Line $I_y = I_R - I_B = -20 < 0 -5 < -60 = -22.5 + j4.33$

$$= 22.91 < 169A$$

B- Line $I_b = I_Y - I_R = 10 < 60 - 20 < 0 = 25 + j8.6 = 26.45 < 19.1A$

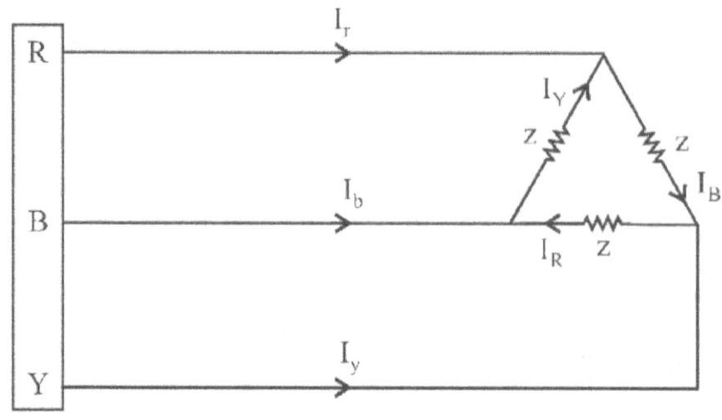

Fig.4.6

4.4 Sequence Impedances and Sequence Networks

1. Introduction

There are three types of sequence impedances of a network or a component in a power system. These are:

- Positive Sequence Impedance
- Negative Sequence Impedance, and
- Zero Sequence Impedance

The impedance for which the flow of positive sequence current is seen in the network is known as positive sequence impedance.

The impedance due to which negative sequence current flows through the network is termed as negative sequence impedance.

The impedance offered to zero sequence currents is called zero sequence impedances of the network.

Let us consider:

Z_{an} = impedance of the load between phase a and neutral n.

Z_{bn} = impedance of the load between phase b and neutral n.

Z_{cn} = impedance of the load between phase c and neutral n.

Therefore, Positive Sequence Impedance of the network is:

$$Z_p = \frac{1}{3}(Z_{an} + aZ_{bn} + a^2Z_{cn})$$

Negative Sequence Impedance of the network is:

$$Z_n = \frac{1}{3}(Z_{an} + a^2Z_{bn} + aZ_{cn})$$

Zero Sequence Impedance of the network is:

$$Z_0 = \frac{1}{3}(Z_{an} + Z_{bn} + Z_{cn})$$

For a three phase symmetrical static circuit which does not have any internal voltages like transformers and transmission lines, the impedances offered to the currents on any phase are equivalent in all the phases and also there is no mutual coupling between the sequence networks.

As in the static devices the sequence has no significance, so the positive and the negative sequence impedances are equal. The zero sequence impedances which includes the impedance of the return path through the ground is usually different from the positive and negative sequence impedances.

In rotating machines, the impedances of the positive sequence components of current and the negative sequence components of currents are different.

By sequence network we mean impedances to current of any one particular sequence only. Hence corresponding to positive, negative and zero sequence currents we have positive, negative and zero sequence networks.

Thus for every power system three sequence network can be formed from these sequence network currents I_{a1}, I_{a2} and I_{a0} and then they can be interconnected in different ways to represent different unbalanced fault conditions.

The sequence currents and voltages during the fault condition are then calculated from which actual fault currents and voltages can then be determined.

Symmetrical faults can be analyzed with the help of positive sequence network. The positive sequence network is nearly same as the impedance or reactance diagram.

Though the negative sequence impedance is also similar to positive sequence network but varies as there is normally no negative sequence emf sources and also negative sequence impedances of rotating machines are generally different from their positive sequence impedances.

The phase displacement of the transformer bank for the negative sequence is of the opposite sign to that of the positive sequence impedances.

The zero sequence network likewise will be free of internal voltages, the flow of current being caused by the voltage at the fault point. The impedance to zero sequence currents are very frequently different from the positive or negative sequence currents.

Zero sequence reactance of the transmission lines is higher than that for positive sequence reactance. The impedance of transformer and generator will depend upon the type of connections (delta or star or grounded or isolated).

2. Synchronous Machines

An unloaded synchronous machine grounded through a reactor of impedance Zn is shown in Fig.4.7.

Here Ea, Eb and Ec are the induced emfs in the three phases. When an unsymmetrical fault occurs on the machine terminals, unbalanced current I_a, I_b and I_c flows in the lines. If fault involves ground, a current I_n flows to neutral from ground by reactor Zn.

The positive sequence impedance and network of a synchronous machine is given in Fig.4.8.

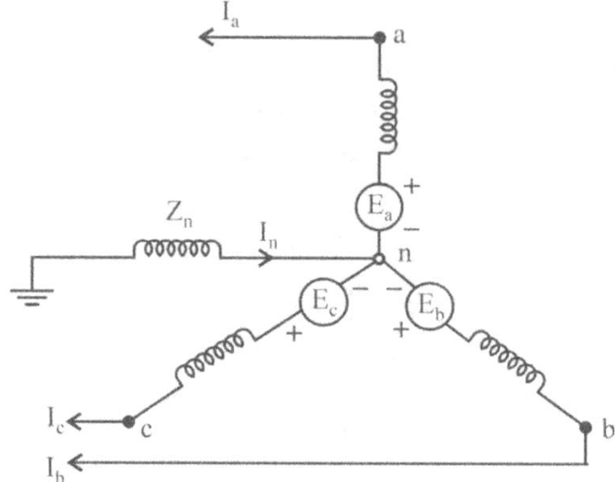

Fig.4.7 Three phase Unloaded Synchronous Generator with Grounded Neutral Through Impedance Zn.

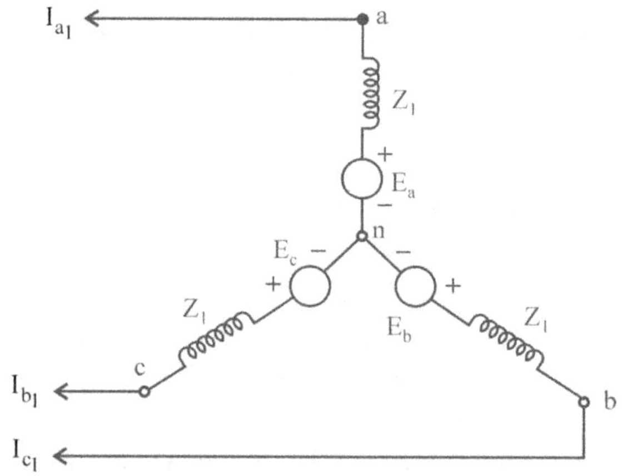

Fig.4.8 Three–Phase Positive Sequence Network Model.

$$V_{a1} = E_a - I_{a1}Z_1$$

The Negative sequence impedance and network of a synchronous machine is given in (Fig. 4.9).

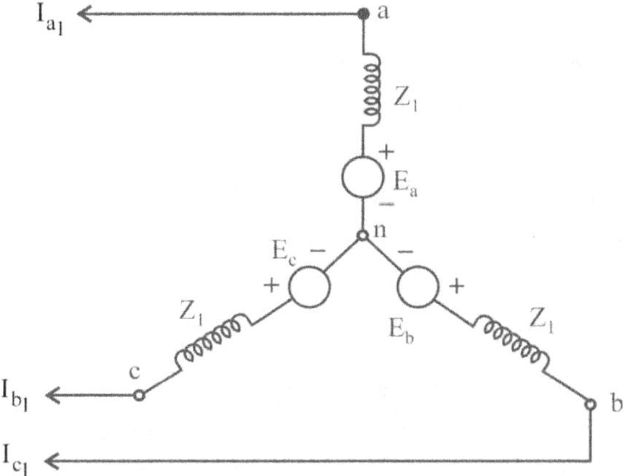

Fig.4.9 Three–Phase Negative Sequence Network Model

$$V_{a2} = -I_{a2}Z_2$$

The Zero sequence impedance and network of a synchronous machine is given in Fig. (4.10).

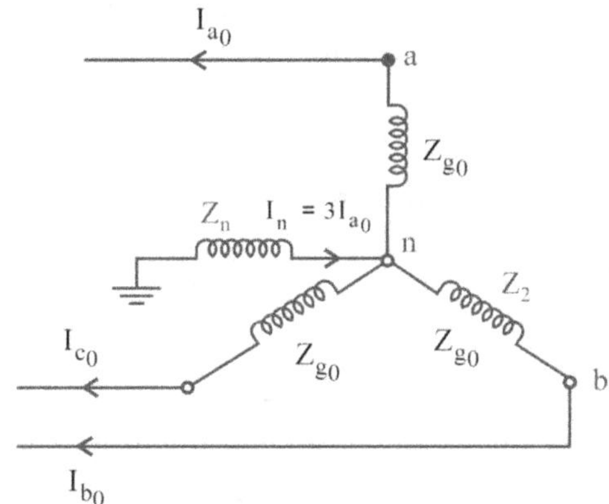

Fig.4.10 Three–Phase Zero Sequence Network Model.

$$Z_0 = 3Z_n + Z_{g0}$$

$$V_{a0} = -I_{a0}Z_0 = -I_{a0}(Z_{g0} + 3Z_n)$$

3. Transmission Lines

A fully transposed three-phase line is completely symmetrical and therefore the positive and negative sequence impedances of a transmission line are independent of phase sequence and are equal. When only zero sequence currents flow in a transmission line, the currents in each phase are identical in both magnitude and phase. Such currents return partly through ground and the rest through overhead ground wires. The zero sequence impedance is about 2 to 4 times the positive sequence impedance.

4. Transformers

Since the resistance of the windings of a transformer is usually small in comparison to the leakage reactance, thus the positive sequence series impedance of a transformer is almost equal to its leakage reactance.

As transformer is a static device, the impedances become independent of the phase order provided that the applied voltages are balanced, thus the positive and the negative sequence impedances are identical.

Thus for a transformer we can write:

$$Z_1 = Z_2 = Z_{\text{leakage}}$$

When three phase transformers are used the situation becomes more complex with regard to zero sequence impedance because there comes more probability of various ways of connecting the transformer.

Assuming the transformer connection such that zero sequence current flows through both the sides of the transformer. The zero sequence impedance offered by the transformer is very slightly different from the positive or negative sequence impedances and thus all are considered equal.

However the zero sequence currents can flow through the windings of the star connection only, if that point is grounded plus the zero sequence current cannot flow in the windings if the star point is isolated.

In case of delta connected windings no zero sequence currents can flow through it as there is no return path available for the zero sequence currents. But if any zero sequence voltage is induced in the delta connected windings, then zero sequence current can flow through the windings, this has been shown in Fig.(4.11).

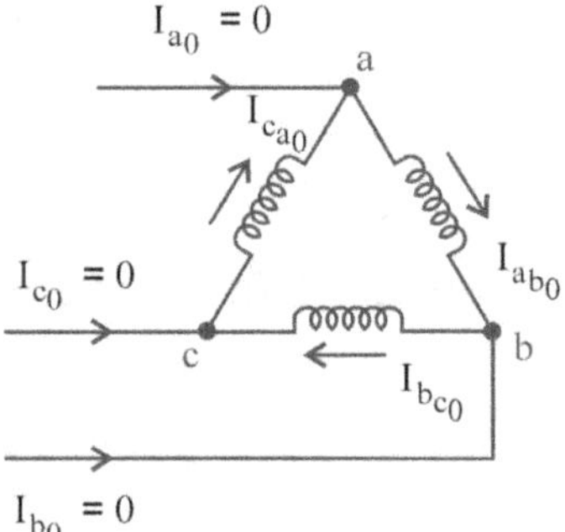

Fig.4.11 Flow of Zero Sequence Current in Delta-Connected Windings.

The determination of various zero sequence network of a transformer can be done from the general circuit (Fig. 4.12).

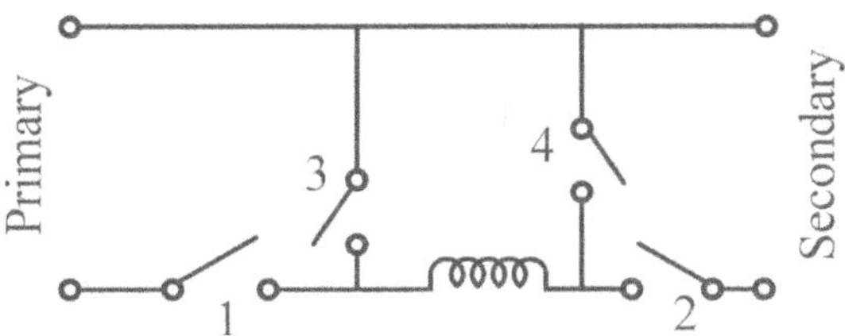

Fig. 4.12 General Circuit For the Determination of
Zero Sequence Network of a Transformer.

Fig. 4.12 shows the general circuit which has Zero Sequence Impedance (Z_0) at the windings of the transformer. There exist 2 series and two shunt switches (one shunt and one series switch for each side).

The series switch of a particular side is closed if its star connection is grounded and the shunt switch is closed if that side has delta connection, otherwise they are left open. The zero sequence representations of transformers for various winding arrangements are given below:

1. **Star- Star transformers with isolated neutrals:** All the four switches are kept open in this connection, as the neutrals of both the primary and secondary side of the transformer are isolated, so zero sequence current cannot flow. The connection diagram with Zero Sequence Equivalent Circuit is given below.

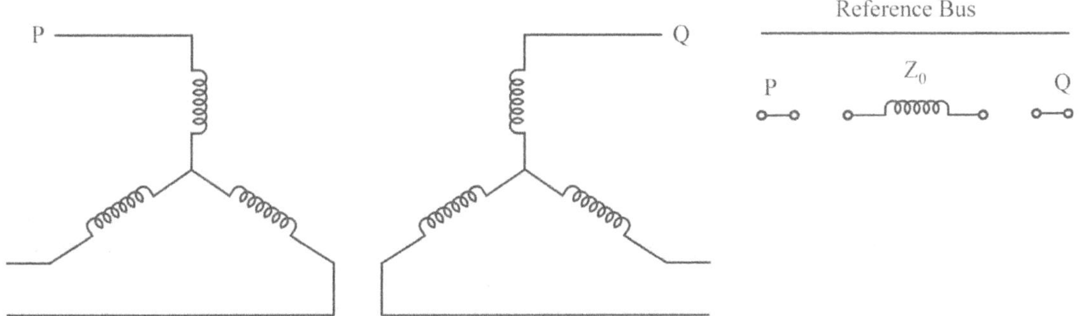

2. **Star–Star transformer with anyone Neutral Grounded:** Since one of the star is ungrounded, thus the zero sequence current cannot flow through the other star which is grounded. Hence open circuit exists between the P and Q (which here is a zero sequence network), only switch 1 is closed rest are all open. The connection diagram with Zero Sequence Equivalent Circuit is given below.

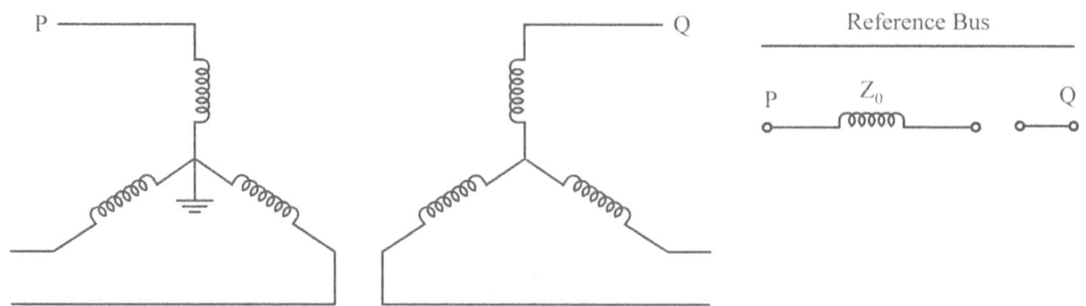

3. **Star–Star transformer with both Neutral Grounded:** As both neutrals of a star-star transformer are grounded, there exists a path in the transformer for the flow of zero sequence currents in both the windings by the grounded neutral. The zero sequence of P and Q are connected by the zero sequence impedance (Z_0) of the transformer. Here switches 1 and 2 are closed, these are the series switches and switch 3 and 4 are kept open, these are the shunt switches.

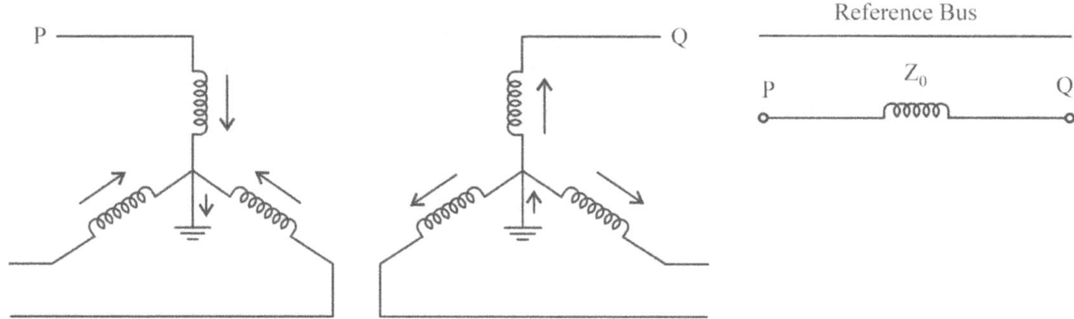

4. **Star–Delta Transformer with grounded star neutral:** Here neutral of the star side is grounded so zero sequence current flows through the star side, hence balancing zero sequence current flows through the delta side. But no zero sequence current can flow in the lines on the delta side. Series switch 1 and shunt switch 4 are closed while series switch 2 and shunt switch 3 are kept open. This means that zero sequence network has path from P through the zero sequence impedance of the transformer to the reference bus on the star side of the transformer, while open circuit exists on line Q on the delta side of the transformer. The connection diagram with Zero Sequence Equivalent Circuit is given below.

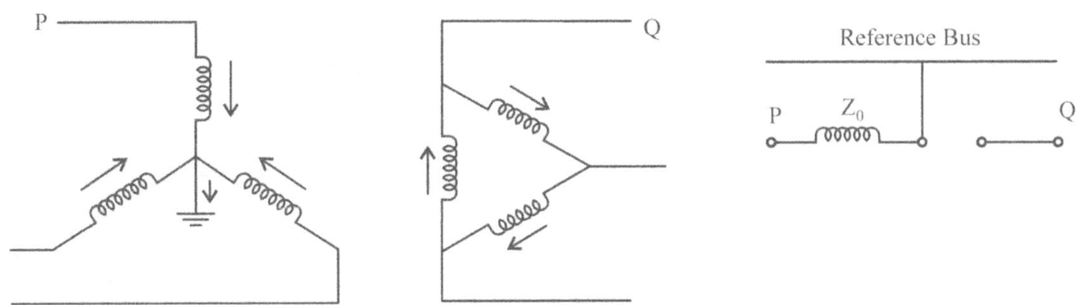

5. **Star–Delta Transformer with grounded star neutral having reactance:** When the star side of the transformer is grounded with reactance Z_n, then an impedance $3Z_n$ appears in series with Z_0 in the sequence network. The connection diagram with Zero Sequence Equivalent Circuit is given below.

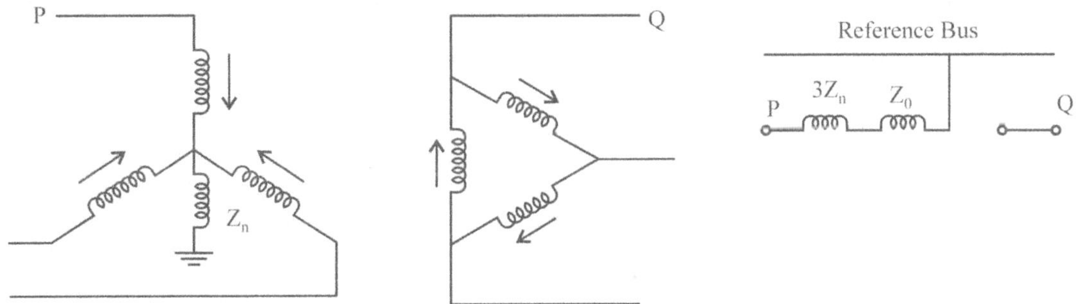

6. **Star–Delta Transformer with isolated star:** As the primary side of the transformer is star connected with neutral isolated, switch 1 and 3 are kept open and the secondary side is delta connected, the shunt switch 4 is closed and the series switch 2 is left open. In this arrangement no zero sequence current can flow in the transformer windings. The connection diagram with Zero Sequence Equivalent Circuit is given below.

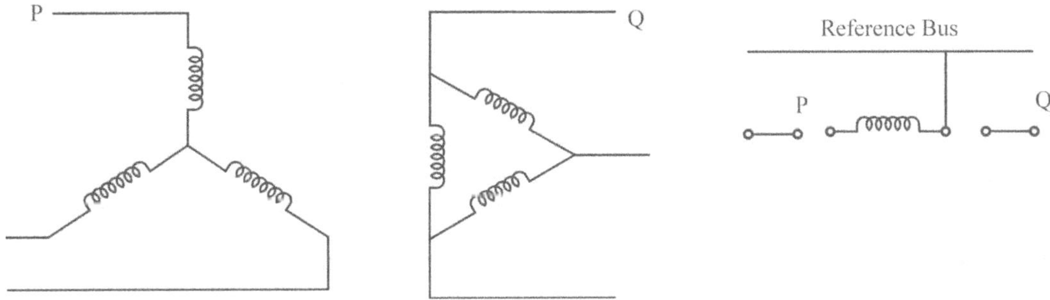

7. **Delta–Delta Transformer:** As both sides are delta connected, so both series switches 1 and 2 are kept open and both shunt switches 3 and 4 are closed. Delta connection does not allow return path to the flow of zero sequence current, thus it cannot flow through delta–delta transformer. However, delta–delta transformer connection allows the circulation of zero sequence current in the delta windings. The zero sequence impedance

is connected to the reference bus on both sides to account for the circulating current in the two deltas. The connection diagram with Zero Sequence Equivalent Circuit is given below.

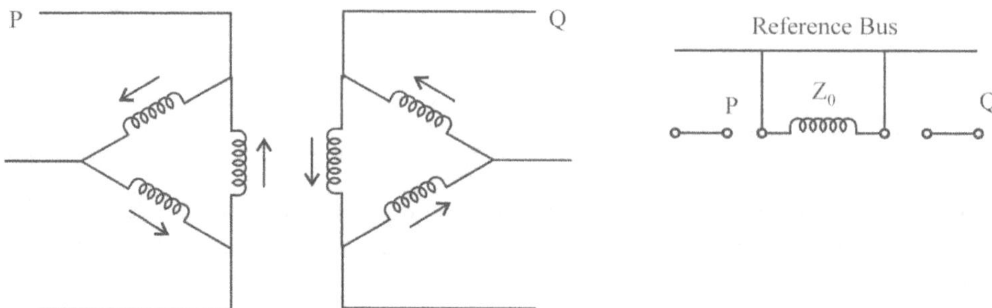

8. **Star–Star with any one Neutral Grounded and Tertiary:** Here the tertiary windings provides path for the zero sequence currents. The connection diagram with Zero Sequence Equivalent Circuit is given below.

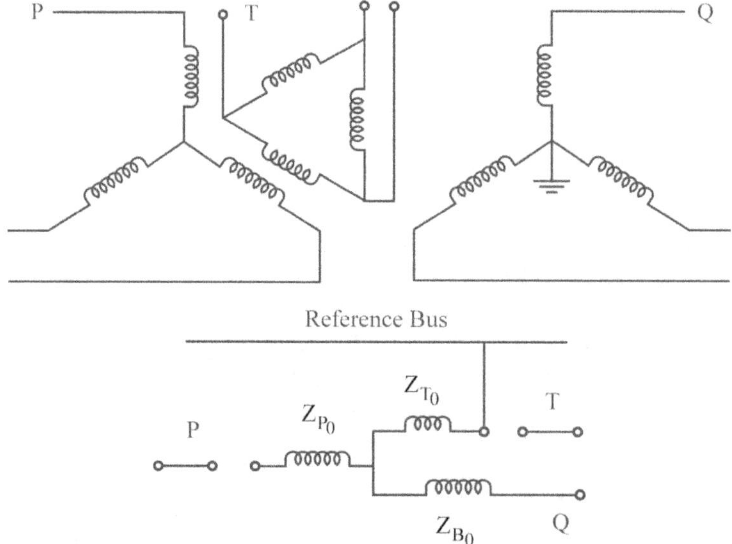

Exercise

Short Questions

1. For a fully transposed transmission line
   a) positive, negative and zero sequence impedances are equal.
   b) positive and negative sequence impedances are equal.
   c) zero and positive sequence impedances are equal.
   d) negative and zero sequence impedances are equal

2. Symmetrical components are used in power system for the analysis of

   a) Balanced 3-phase fault

   b) Unbalanced 3-phase fault

   c) Normal power system under steady conditions

   d) Stability of system under disturbance

3. If the positive, negative and zero-sequence reactance of an element of a power system are 0.3, 0.3 and 0.8 p.u. respectively, then the element would be a?

   a) Synchronous generator

   b) Synchronous motor

   c) Static load

   d) Transmission line

Board Questions

1. Describe the concept of 'a' operator.

2. How do you calculate sequence impedance?

3. What is the importance of sequence impedances?

4. Why are positive and negative sequence impedances same?

5. What is the practical application of sequence impedances?

6. What do you understand by symmetrical component analysis? Explain its analytical methods.

7. A 3-phase transmission line operating at 33kV and having a resistance of 0.15 Ω and reactance of 1.12Ω is connected to the generating station bus-bars through 5MVA step-up transformer having a reactance of 0.02 p.u. The bus-bars are supplied by a10MVA alternator having 0.068p.u. reactance. A load of 25HP motor is running at a p.f. 0.866, Calculate the short-circuit kVA fed to symmetrical fault between phases if it occurs (i) at the load end of transmission line (ii) at the high voltage terminals of the transformer.

8. The section bus-bars A and B are linked by a bus-bar reactor rated at 6000 kVA with 0.2p.u. reactance. On bus-bar A, there are two generators each of 14,500 kVA with 0.158 p.u. reactance and on B two generators each of 5000 kVA with 0.056 p.u. reactance. Find the steady MVA fed into a dead short circuit between all phases on B with bus-bar reactor in the circuit.

9. Find the original unbalanced phaser's for the following symmetrical components of a set of unbalanced three phase currents

$$I_{a0} = 4 \angle -40°A, \; I_{a1} = 6 \angle 100°A, \; I_{a2} = 5 \angle 40°A$$

10. A Delta connected load of resistors of 12Ω, 15Ω and 222Ω are supplied power by a balanced 440 volts supply. Determine the sequence components currents in the resistors and in supply lines?

11. A $3\varnothing$, 4-wire system is operating under unbalance condition. During the unbalancing it was observed that R phase positive sequence and the negative sequence component were $120\angle 0$A and $260 < 72$ A respectively. It was also observed that $440 < 360$. A current flowing back to the supply through neutral conductor. Determine the current in the three lines.

12. Figure shows the current is flowing to the Δ-connected load from a 3 phase supply. B phase terminal is open; determine the symmetrical components of the line currents.

Chapter 5

Asymmetrical Faults in Power Systems

5.1 Introduction

Faults are power system hazard. In many cases, it can harm the device. This causes the unexpected magnitude of voltages and currents in the systems. It can damage the insulation and therefore decrease the equipment's life. It leads to system instability. Thus, it is necessary to identify and avoid the likelihood of any fault.

Unsymmetrical faults in the power system are the most common type of faults. For this kind of faults, in the three phase lines, they corresponds to unsymmetrical currents of unequal magnitude and unequal phase displacement. Unsymmetrical faults involve either one or two phases whereas for symmetrical faults affecting all three phases.

Following assumptions have to be considered during the study of unsymmetrical faults are:

a. During balanced condition only positive sequence is active.

b. Heavy current only flows through the faulty line while load current of the other healthy phase line current is negligible.

c. Fault impedance and transmission line charging capacitance are neglected and considered as zero.

d. R-phase considered as reference phase,

The fault currents will be represented as I_R, I_Y, I_B and the electromotive forces per phases are presented by E_R, E_Y, E_B and the line voltages will be represented by V_R, V_Y, V_B.

5.2 Single Line-To-Ground Fault

Let's take a 3-phase power system for the study of a single line to ground fault. There is a line-to-ground fault on the system's red phase.

Now, we know that, fault occurred at R-phase. Thus the other phase currents such as I_Y, I_B are very small in magnitude compared to faulty phase current I_R and R phase is connected to ground. Thus R phase terminal potential is same as ground potential. Thus, mathematically we can write, $I_Y = I_B = 0$ and $V_R = 0$.

$I_0 = 1/3(I_R + I_Y + I_B) = 1/3\ I_R$

$I_1 = 1/3(I_R + a\ I_Y + a^2\ I_B) = 1/3\ I_R$

$I_2 = 1/3(I_R + a^2\ I_Y + a\ I_B) = 1/3\ I_R$

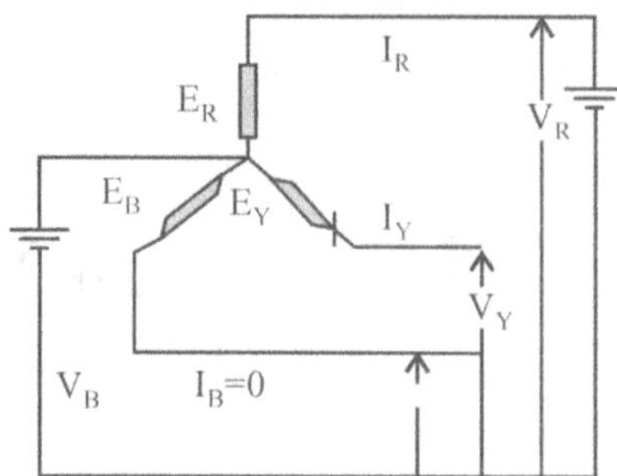

Fig.5.1

Therefore,

$I_0 = I_1 = I_2 = 1/3 \, I_R$

For the analysis of the fault, firstly need to derive the fault current.

Let us consider the fault loop with the fault line and ground.

$E_R = I_0 Z_0 + I_1 Z_1 + I_2 Z_2 + V_R$

Since we know that $V_R = 0$ and $I_1 = I_2 = I_0$

Therefore, $I_0 = E_R/(Z_0 + Z_1 + Z_2)$

Since, $I_0 = I_R/3$

Therefore, $I_R = 3E_R/(Z_0 + Z_1 + Z_2)$

The above expression gives the fault current flowing through the circuit.

Using this fault current we can draw the equivalent circuit. From the expression of fault current, it is clear that the equivalent circuit will be as follows,

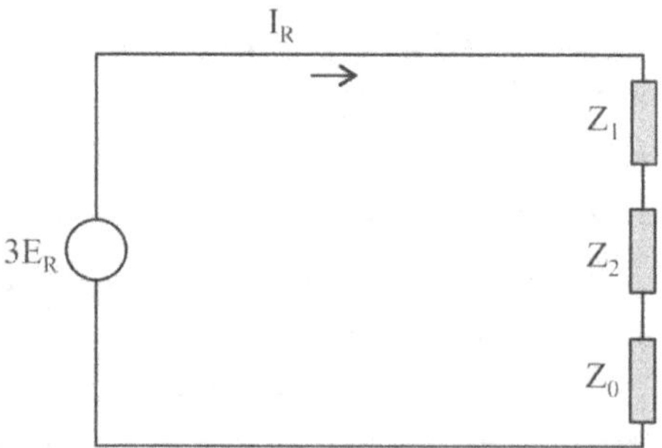

Fig.5.2 Equivalent Circuit.

Making such equivalent circuit makes calculations easier and analysis simpler.

Further on let us evaluate the phase voltages at the location of fault. The voltage between line and the fault will give us the same.

We know that the generated electro motive force is of positive sequence only, therefore the sequence voltages at the fault will be given by,

$V_0 = -E_R Z_0/(Z_0 + Z_1 + Z_2)$

$V_1 = E_R - I_1 Z_1 = E_R (Z_2 + Z_0)/(Z_0 + Z_1 + Z_2)$

$V_2 = -E_R Z_2/(Z_0 + Z_1 + Z_2)$

Hence the phase voltages at fault will be given by:

$V_R = 0$

$V_Y = V_0 + a^2 V_1 + a V_2$

$V_B = V_0 + a V_1 + a_2 V_2$

Problems

Example 1: A 3-phase, 13.2 kV, 30 MVA generator with $X_0 = 0.06$ p.u., $X_1 = 0.24$ p.u. and $X_2 = 0.24$ p.u. is grounded through a reactance of 0.36Ω. Calculate the fault current for a single line to ground fault.

Solution:

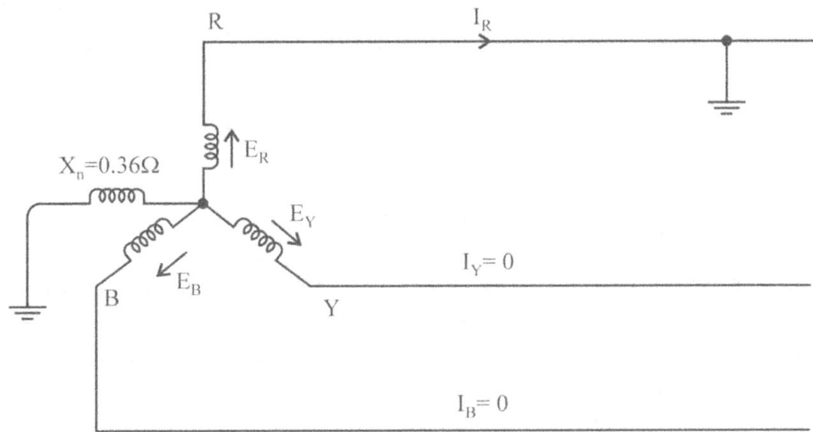

Fig.5.3

The fault is assumed to occur on the red phase. Let's take the red phase and its e.m.f. be $E_R = 1$p.u.

The emf and other parameter are in p.u.so reactance also should have the valve in p.u..

$$\text{p. u. of } X_n = \frac{\text{KVA rating}}{(kv \times kv)} \times X_n \text{ in } \Omega$$

$$= 0.36 \times \frac{30.000}{13.2 \times 13.2 \times 1000}$$

$$= 0.0619 \text{ p.u.}$$

For the line to ground fault we have,

$$I_1 = I_2 = I_0 = \frac{E_R}{x_1 + x_2 + x_0 + 3 \times x_n}$$

$$= \frac{1}{j0.24 + j0.24 + j0.06 + 3 \times 0.0619}$$

$$= \frac{1}{0.72595} = -j1.3775 \text{ p.u.}$$

Fault current $I_R = 3I_0 = 3 \times (-1.3775)$

$$= 4.1325 \text{p.u.}$$

Fault current in amperes = rated current $\times$ p.u. value

$$= 4.1325 \times [30 \times 10^6 / \sqrt{3} \times 13.2 \times 1000$$

$$= 5422.5 \text{ A}$$

Example 2: A 3-phase, 13.2 kV, 12 MVA alternator has sequence reactances of $X_0 = 0\cdot06$ p.u., $X_1 = 0\cdot18$ p.u. and $X_2 = 0\cdot18$ p.u. If the generator is on no load, find the ratio of fault currents for L-G fault to that when all the 3-phases are dead short-circuited.

Solution:

By taking the red phase as the reference. Let its phase e.m.f. be $E_R = 1$p.u.

 Line to ground fault:

Suppose the fault current occurs on the red phase. Then,

$$I_1 = I_2 = I_0 = \frac{E_R}{x_1 + x_2 + x_3}$$

$$I_0 = \frac{1}{j0.18 + j0.18 + j0.06}$$

$$= -j2.380$$

So the fault current is $I_R = 3 \times I_0$

$$= 3 \times (-j2.380)$$

$$= -j7.1428$$

Three Phase Fault

When a dead short circuit occurs on all the three phases, it gives rise to symmetrical fault currents. Therefore, the fault current (say I_{sh}) is limited by the positive sequence reactance only.

Fault current, $I_{sh} = E_R/X_R = \dfrac{1}{j0.18}$

$$= -j5.55$$

Ratio of the fault current $= I_R/I_{sh} = \dfrac{-j7.1428}{-j5.55}$

$$= 1.285$$

So we can say that single line-to-ground fault current is $1 \cdot 285$ times that due to dead short circuit on the 3-phases.

5.3 Line-To-Line Fault

Let's take a 3-phase power system for the study of a line-to line fault. Consider that there is a line-to-line fault between the system's Blue and Yellow phase.

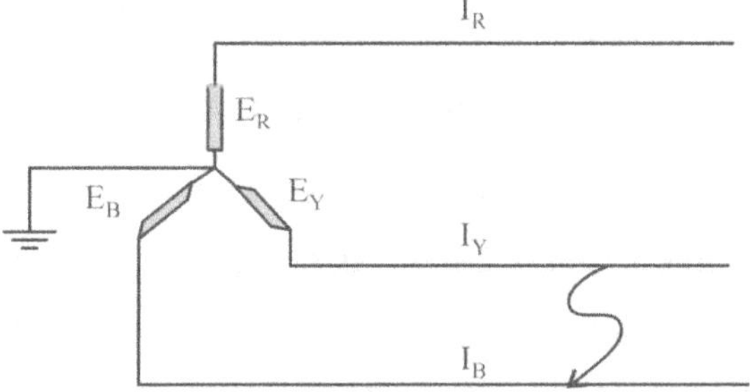

Fig.5.4

Now, since a fault has occurred between phase Y and B, the B-Y phase terminal potential will be same, and the phase current I_R will be very small in magnitude compared to faulty phase current $I_B,$ I_Y and I_{BY}. Thus, mathematically we can write,

$V_Y = V_B,$ $I_R = 0$ and $I_Y + I_B = 0$

Now,

$V_Y = V_B$

$V_0 + a^2 V_1 + a V_2 = V_0 + a V_1 + a^2 V_2$

$V_1 = V_2$

We also know that,

$I_Y + I_B = 0$

On substituting the sequence values we get,

$I_1 + I_2 = 0$

For the analysis of the fault system, we will first derive the fault current,

Fault current $I_Y = I_0 + a^2 I_1 + a I_2$

$= 0 + a^2 (E_R/(Z_1+Z_2)) + a(E_R/(Z_1+Z_2))$

$= -j\sqrt{3} E_R/(Z_1+Z_2)$

$= -I_B$

Based on the fault current value, we can determine the equivalent circuit. The equivalent circuit helps in easier calculation and analysis.

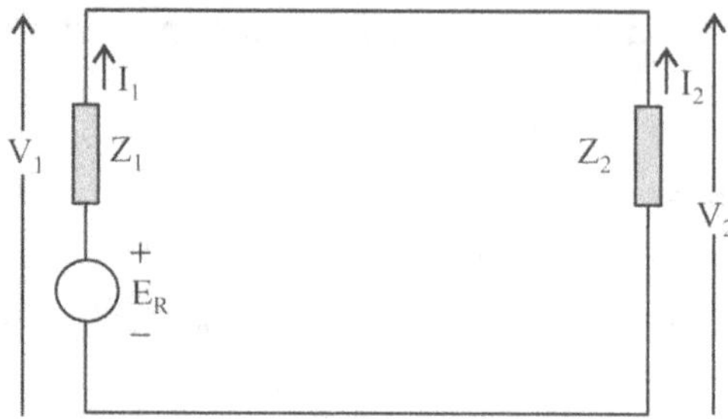

Fig.5.5 Equivalent Circuit.

Further on let us evaluate the phase voltages,

Since the electro motive force has only the positive sequence component, therefore

$E_0 = 0, E_2 = 0, E_1 = E_R$

The sequence voltage components will be given by,

$V_0 = 0$

$V_1 = E_R Z_2/(Z_1 + Z_2)$

$V_2 = E_R Z_2/(Z_1 + Z_2)$

Further the phase voltages at the point of fault will be given by,

$V_R = V_0 + V_1 + V_2 = 2Z_2 E_R/(Z_1 + Z_2)$

$V_Y = -Z_2 E_R/(Z_1 + Z_2)$

$V_B = -Z_2 E_R/(Z_1 + Z_2)$

5.4 Double Line-to-Ground Fault

Let's take a 3-phase power system for the study of a Double line-to ground fault. There is a double line-to-ground fault between the system's Blue and Yellow phase and the ground.

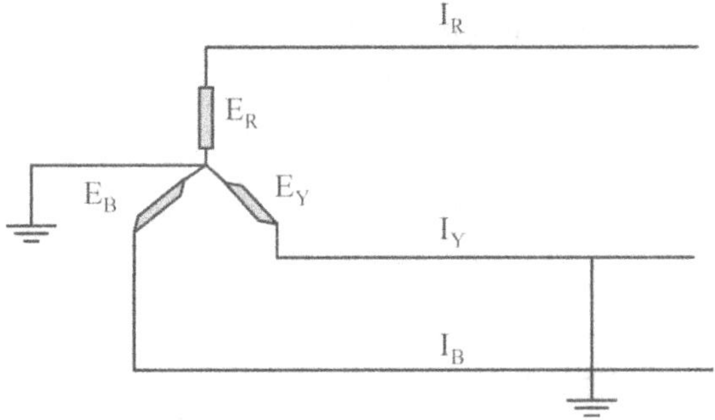

Fig.5.6

Now, we know that, fault occurred between Y-B phases and ground. Thus the other phase current such as I_R is very small in magnitude compared to faulty phase current $I_{B.}$, I_Y and B, Y phases are together with ground connectivity. Thus B, Y phase terminal potential is same as ground potential. Thus, mathematically we can write,

$I_R = 0$, $V_Y = V_B = 0$,

Further on it can be understood that, $V_1 = V_2 = V_0 = 1/3\ V_R$

and, $I_R = I_1 + I_2 + I_0 = 0$

Now, for the analysis of the fault system we will derive the fault current, it can be incurred that,

$I_1 = E_R / ((Z_1 + (Z_2\ Z_0\ (Z_2 + Z_0))$

$I_0 = -I_1\ Z_2\ (Z_1 + Z_2)$

Therefore, the fault currents are given by,

$I_Y + I_B = I_0 = -3Z_2\ E_R / (Z_0 Z_1 + Z_0 Z_2 + Z_1 Z_2)$

From the above fault expression, we can determine the equivalent circuit.

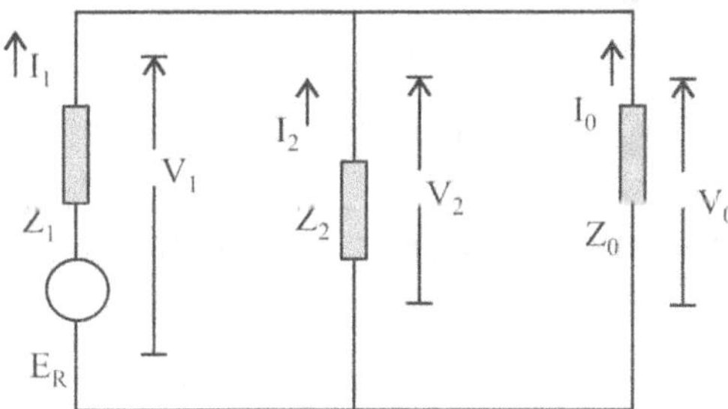

Fig.5.7 Equivalent Circuit.

For further analysis, let us also evaluate the phase voltages.

The sequence voltages of phase R will be given by,

$V_1 = V_2 = V_0 = 1/3 V_R$

and, $V_1 = E_R - I_1 Z_1$

$V_2 = -I_2 Z_2$

$V_0 = -I_0 Z_0$

Therefore, the phase voltages will be,

$V_R = 3 V_2$

$V_Y = 0$

$V_B = 0$

Problems

Example 1: If in a system, the fault is a double line-to-ground, and if the per unit values of the reactance of the network of the 3 lines at fault are given as follow, $Z_1 = j\ 0.2$ p.u., $Z_2 = j\ 0.1$ p.u., $Z_0 = j\ 0.09$ p.u. Find the fault current of the double line-to-ground using the given data.

Solution:

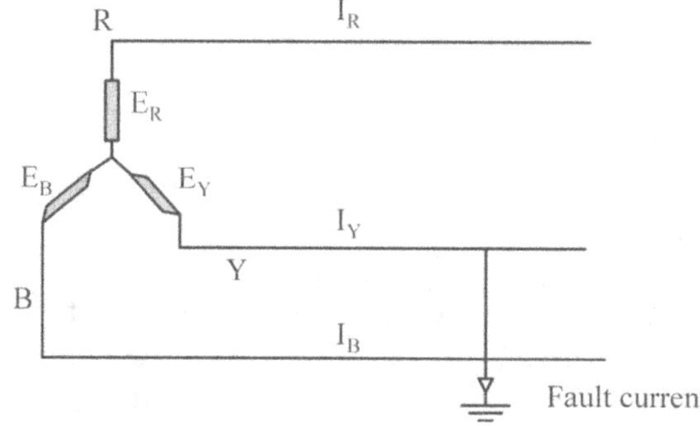

Fig.5.8

Here, if the faults are at Y and B and ground R as reference, we can say that the phase emf of it be $E_R = 1$p.u.

Given,

$Z_1 = j\ 0.2$ p.u.

$Z_2 = j\ 0.1$ p.u.

$Z_0 = j\ 0.09$ p.u.

Fault current is given by,

$I_F = I_Y + I_B$

$$= -3 \times (Z_2 \, E_R/Z_1 \, Z_2 + Z_1 \, Z_0 + Z_2 \, Z_0)$$

$$= -3 \times j \, 0.1 \times 1/\{(j0.2 \times j0.1) + (j0.2 \times j0.09) + (j \, 0.1 \times j0.09)\}$$

$$= -j \, 0.3/0.047$$

$$= j \, 6.382 \text{ p.u.} \quad \text{(This is the fault current)}$$

Example 2: In a system, the per unit values of positive, negative and zero sequence reactance of a network at fault are 0·08, 0·07 and 0·05. Determine the fault current if the fault is double line-to-ground.

Solution: Let us consider that the fault involves Y and B and the ground taking R as the reference phase and let the phase emf of it be $E_R = 1$p.u.

Now, $X_1 = j \, 0.112$ p.u.

$X_2 = j \, 0.098$ p.u.

$X_0 = j \, 0.07$ p.u.

Now we know that for double line-to-ground fault, we have, Fault current,

$I_F = I_Y + I_B$

$$= -3x_2 \, E_R/X_1 \, X_2 + X_1 \, X_0 + X_2 \, X_0$$

$$= -3 \times j \, 0.098 \times 1/\{(0 \, j2.0112 \times j0.098) + (j0.112 \times j0.07) + (j \, 0.098 \times j0.07)\}$$

$$= -j0.294/0.025676$$

$$= j11.45 \text{ p.u.}$$

Exercise

Short Questions

1. What is the value of zero sequence impedance in line to line faults?

   a) $Z_0 = 1$ _______________________________ b) $Z_0 = \infty$

   c) $Z_0 = 3 \, Z_n$ _______________________________ d) $Z_0 = 0$

2. What percentage of fault occurring in the power system is line to line fault?

   a) 5 %   b) 30 %

   c) 25 %   d) 15 %

3. What happens to the value of the fault current in case of SLG fault, if fault impedance is introduced?

    a) The fault current increase

    b) The fault current remains same as in case of SLG fault.

    c) The fault current becomes zero

    d) The fault current is reduced

Board Questions

1. Which type of asymmetrical faults is most severe?

2. Which condition gives rise to the most severe fault condition?

3. How does circuit current behave on the occurrence of fault in a power system?

4. What is an unsymmetrical fault?

5. What do you understand by positive, negative and zero sequence impedances? Discuss them with reference to synchronous generators, transformers and transm

6. Derive an expression for fault current for single line to ground fault by symmetrical component method.

7. Derive an expression for fault current for line to line fault by symmetrical component method.

8. Derive an expression for fault current for Double line to ground fault by symmetrical component method.

9. A 3-phase, 13.2 kV, 15 MVA alternator has sequence reactance's of $X_0 = 0\cdot02$ p.u., $X_1 = 0\cdot15$ p.u. and $X_2 = 0\cdot20$ p.u. If the generator is on no load, find the ratio of fault currents for L-G fault to that when all the 3-phases are dead short-circuited.

10. In a system, the per unit values of positive, negative and zero sequence reactance of a network at fault are $0\cdot06$, $0\cdot08$ and $0\cdot02$. Determine the fault current if the fault is double line-to-ground.

11. A 3-phase, 11 kV, 15 MVA generator with $X_0 = 0\cdot02$ p.u., $X_1 = 0\cdot12$ p.u. and $X_2 = 0012$ p.u. is grounded through a reactance of $0\cdot36\Omega$. Calculate the fault current for a single line to ground fault.

Answers

1. d	2. d	3. d

Chapter 6

Load Flow Analysis

6.1 Introduction

The key purpose of a Load Flow is to find out the flow of active and reactive powers all over the network and the voltage levels and the phase angles at every bus of the system at the steady state operating conditions. The Power System expansion and its planning as well as the daily operation of power system depend on with full trust on numerous power flow studies. The power engineer checked the recommended operating limit of nodal voltage magnitudes, active and reactive power flows on transmission lines through the information suggested by Load Flow studies. Subsequently, the power engineer also takes corrective measure if the power flow results are beyond of operating limits.

Classification of Different Type of Bus and Power Flow Equation

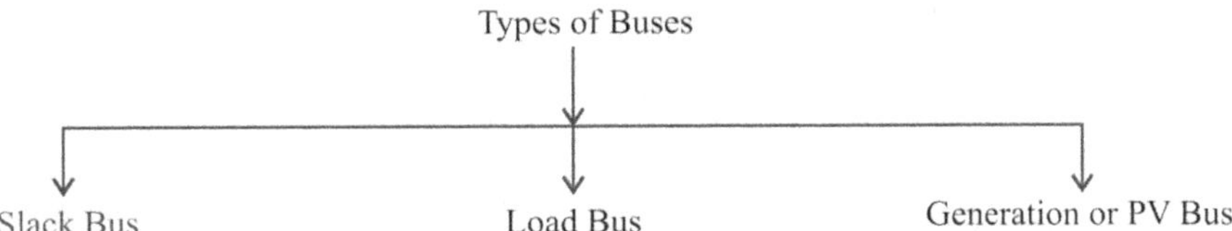

6.2 Classification of Bus

A bus is a node at which one or several network element such as transmission lines, loads and generators are joined. There may not be connected all of these at a single node. There are four quantities associated with each bus, such as magnitude of voltage $|V|$, phase angle of voltage δ, active power P and reactive power Q. In load flow analysis two quantities are specified and other two need to be found out.

Bus Classification

Type of Bus	Given quantities	Unknown quantities
Generation or P-V Bus	$P, |V|$	Q, δ
Load or P-Q Bus	P, Q	$|V|, \delta$
Slack Bus or Reference Bus or Swing Bus	$|V|, \delta$	P, Q

- **Generation Bus** or **P-V Bus:** This is also known as voltage controlled bus. In this bus the magnitude of generation voltage and active power are specified to its rating is given. The unknown reactive power generation and phase angle of the voltage generation are to be calculated.

- **Load Bus:** In this bus total injected power is given. Unknown magnitude and phase angle of voltage are to be calculated.

- **Slack Bus or Reference Bus or Swing Bus:** For this kind of bus, the magnitude and the phase angle of voltage generation information are known and the other two quantities need to be calculated i.e. active and reactive power. Normally the phase angle is considered as zero.

6.3 Power Flow Equation

The complex power injected by the generating source into the ith bus of a power system is given by

$$S_i = P_i + jQ_i = V_i I_i^* \text{ where } i = 1, 2, 3.....n \qquad .....(6.1)$$

Here, V_i is the voltage at the Ith bus with respect to ground and I_i^* is the complex conjugate of current of source I^{th} injected into the bus.

Now, we can write

$$S_i^* = P_i - jQ_i = V_i^* \times I_i; \text{ where } i = 1, 2, 3....n \qquad .....(6.2)$$

We know that current into I^{th} bus is expressed by

$$I_i = \Sigma_{k=1}^{n} Y_{iK} V_k \qquad .....(6.3)$$

$$S_i^* = P_i - jQ_i = V_i^* \times \Sigma_{k=1}^{n} Y_{ik} V_k; \text{ where } i=1,2,3,........n \quad .....(6.4)$$

Equating real and imaginary parts we have

Real power $= P_i = R_e\{V_i^* \times \Sigma_{k=1}^{n} Y_{iK} V_k\}$ $\qquad\qquad .....(6.5)$

Reactive power $Q_i = -I_m\{V_i^* \times \Sigma_{k=1}^{n} Y_{iK} V_k\}$ $\qquad\qquad .....(6.6)$

In polar form $\qquad\qquad V_i = |V_i| < \delta_i V_i^* = |V_j| <- \delta_L$ $\qquad\qquad .....(6.7)$

$$Y_{ik} = Y_{ik} < \emptyset_{ik}$$

Thus active and reactive power is expressed as

Real power $= P_i = |V_i| \Sigma_{k=1}^{n} |Y_{iK}||V_k| \cos(\emptyset_{ik} + \delta_k - \delta_i)$ $\qquad\qquad .....(6.8)$

Reactive power $= Q_i = -|V_i| \Sigma_{k=1}^{n} |Y_{iK}||V_k| \sin(\emptyset_{ik} + \delta_k - \delta_i)$ $\qquad\qquad .....(6.9)$

The equation. (6.8) ,(6.9) are known as static load flow equation.

6.4   Admittance Bus Matrix

In an energy system the power is coming to a bus from generators and the loads are taken out power from the bus. The bus may have only generators/loads or both. To show algorithm for finding admittance bus matrix an example of generalized 4 bus network has explained.

Let us assume a 4-bus system as shown in the figure 6.1 below, The impedances of the lines are given in per unit on a common MVA base and for the sake of simplicity it is assumed that there are no resistances in the line.

To find the admittance bus matrix, Kirchhoff's current law has been employed and

Impedances are changed to admittance i.e.

$$y_{ij} = \frac{1}{z_{ij}} = \frac{1}{r_y + jx_{ij}}$$

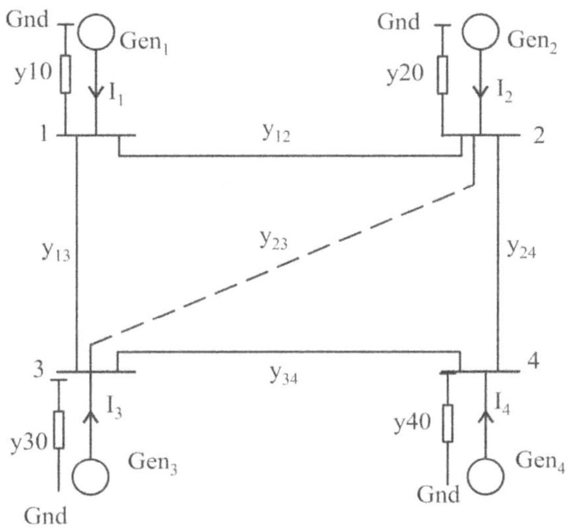

Fig.6.1

Now employing, Kirchhoff's current law to the four different nodes the following equations are obtained:

$$I_1 = v_1 y_{10} + (v_1 - v_2)y_{12} + (v_1 - v_3)y_{13}$$

$$I_2 = v_2 y_{20} + (v_1 - v_2)y_{12} + (v_2 - v_3)y_{23} + (v_2 - v_4)y_{24}$$

$$I_3 = v_3 y_{30} + (v_3 - v_1)y_{13} + (v_3 - v_2)y_{23} + (v_3 - v_4)y_{34}$$

$$I_4 = v_4 y_{40} + (v_4 - v_2)y_{24} + (v_4 - v_3)y_{34}$$

The above equation may be expressed in matrix form as:

$$\begin{bmatrix} I_1 \\ I_2 \\ I_3 \\ I_4 \end{bmatrix} = \begin{bmatrix} y_{10} + y_{12} + y_{13} & -y_{12} & -y_{13} & 0 \\ -y_{12} & y_{20} + y_{12} + y_{23} + y_{24} & -y_{23} & -y_{24} \\ -y_{13} & -y_{23} & y_{30} + y_{13} + y_{23} + y_{34} & -y_{34} \\ 0 & -y_{24} & -y_{34} & y_{40} + y_{24} + y_{34} \end{bmatrix} \begin{bmatrix} v_1 \\ v_2 \\ v_3 \\ v_4 \end{bmatrix}$$

From the matrix it is observed that, the self-admittance terms:

$$Y_{11} = y_{10} + y_{12} + y_{13}$$

$$Y_{22} = y_{20} + y_{12} + y_{23} + y_{24}$$

$$Y_{33} = y_{30} + y_{13} + y_{23} + y_{34}$$

$$Y_{44} = y_{40} + y_{24} + y_{34}$$

From the matrix it is observed that, the mutual admittances

$$Y_{12} = Y_{21} = -y_{12}$$
$$Y_{13} = Y_{31} = -y_{13}$$
$$Y_{14} = Y_{41} = -y_{14} = 0$$
$$Y_{23} = Y_{32} = -y_{23}$$
$$Y_{24} = Y_{42} = -y_{24}$$
$$Y_{34} = Y_{43} = -y_{34}$$

Matrix may be expressed in terms of self-bus admittance (Y_{ii}) and mutual bus admittance (Y_{ik}) :

Example 1: For the network shown in Fig. 6.2, the system data is given in Table 6.1. Find the admittance matrix of the network.

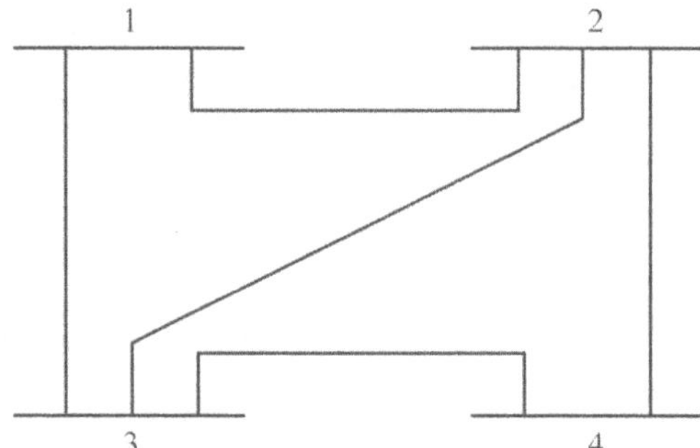

Fig.6.2 One line diagram of 4 Bus System.

Table 6.1 Bus Data.

Bus Data		
BUS	R	X
1-2	0.06	0.26
1-3	0.2	0.4
2-3	0.26	0.56
2-4	0.2	0.4
3-4	0.06	0.26

Here, the number of transmission lines are five. Z_{ij} indicates the impedance of the line which is connected between i and j bus.

$Z_{12} = (0.06 + j0.26); Z_{13} = (0.2 + j0.4)$

$Z_{23} = (0.26 + j0.56); Z_{24} = (0.20 + j0.40); Z_{34} = (0.06 + j0.26)$

Thus,

$y_{12} = 1/(0.06 + j0.26) = 0.84269662 - j3.651685$

$y_{13} = 1/(0.2 + j0.4) = 1 - j2$

$y_{23} = 1/(0.26 + j0.56) = 0.682056 - j1.469045$

$y_{24} = 1/(0.20 + j0.40) = 1 - j2$

$y_{34} = 1/(0.06 + j0.26) = 0.84269662 - j3.651685$

y_{ij} indicates the admittance of the line which is connected between i and j bus.

$Y_{11} = y_{12} + y_{13} = 0.84269662 - j3.651685 + 1 - j2 = 1.84269662 - j5.651685$

$Y_{12} = -y_{12} = -0.8426966 + j3.651685$

$Y_{13} = -y_{13} = -1 + j2$

$Y_{14} = -y_{14} = 0$

$Y_{21} = -y_{12} = -0.8426966 + j3.651685$

$Y_{22} = y_{21} + y_{23} + y_{24} = 0.84269662 - j3.651685 + 0.682056 - j1.469045 + 1 -$
$\quad j2 = 2.52475262 - j7.12073$

$Y_{23} = -y_{23} = -0.682056 + j1.469045$

$Y_{24} = -y_{24} = -1 + j2$

$Y_{31} = -y_{13} = -1 + j2$

$Y_{32} = -y_{23} = -0.682056 + j1.469045$

$Y_{33} = y_{31} + y_{32} + y_{34} = 1 - j2 + 0.682056 - j1.469045 + 0.8426966 - j3.65168$
$\quad = 2.5247526 - j7.120725$

$Y_{34} = -y_{34} = -0.8426966 + j3.65168$

$Y_{41} = -y_{14} = 0$

$Y_{42} = -y_{42} = -1 + j2$

$Y_{43} = -y_{34} = -0.8426966 + j3.65168$

$Y_{44} = y_{41} + y_{42} + y_{43} = 0 + 1 - j2 + 0.8426966 - j3.65168 = 1.8426966 - j5.65168$

$Y_{bus}=$

$$= \begin{bmatrix} 1.8426966 - j5.65168 & -0.8426966 + j3.651685 & -1 + j2 & 0 \\ -0.8426966 + j3.651685 & 2.52475262 - j7.12073 & -0.682056 + j1.469045 & -1 + j2 \\ -1 + j2 & -0.682056 + j1.469045 & 2.5247526 - j7.120725 & -0.8426966 + j3.65168 \\ 0 & -1 + j2 & -0.8426966 + j3.65168 & 1.8426966 - j5.65168 \end{bmatrix}$$

Example 2: For the network shown in Fig.6.3, the system data is given in Table 6.2. Find the admittance matrix of the network.

Table 6.2 Bus DATA.

BUS	R	X
1-2	0.1	0.1
1-3	0.1	0.1
2-3	0.1	0.1

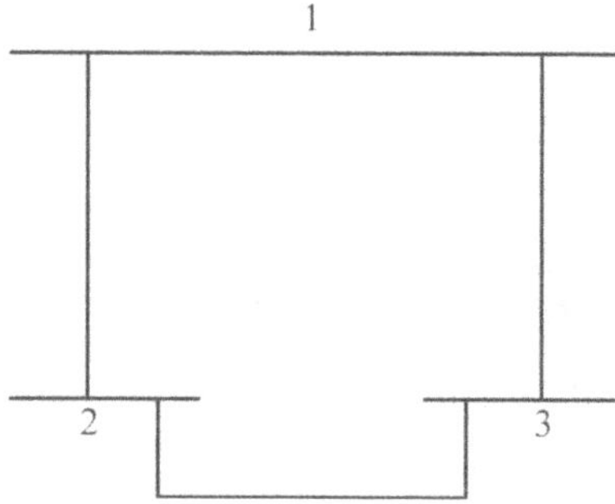

Fig.6.3 One line diagram of 3 Bus System.

Here, the number of transmission lines are three. Z_{ij} indicates the impedance of the line which is connected between i and j bus.

$z_{12} = (0.1 + j0.1)$

$z_{13} = (0.1 + j0.1)$

$z_{23} = (0.1 + j0.1)$

Hence

$y_{12} = 1/(0.1 + j0.1) = 5 - j5$

$y_{13} = 1/(0.1 + j0.1) = 5 - j5$

$y_{23} = 1/(0.1 + j0.1) = 5 - j5$

Y_{ij} indicates the admittance of the line which is connected between i and j bus.

$Y_{11} = y_{12} + y_{13} = 5 - j5 + 5 - j5 = 10 - j10$

$Y_{12} = -y_{12} = -5 + j5$

$Y_{13} = -y_{13} = -5 + j5$

$Y_{21} = -y_{12} = -5 + j5$

$y_{22} = y_{21} + y_{23} = 10 - j10$

$Y_{23} = -y_{23} = -5 + j5$

$Y_{31} = -y_{13} = -5 + j0.5$

$Y_{32} = -y_{23} = -5 + j5$

$Y_{33} = y_{31} + y_{32} = 10 - j10$

Y bus Matrix,

$$Y_{Bus} = \begin{bmatrix} 10 - j10 & -5 + j5 & -5 + j5 \\ -5 + j5 & 10 - j10 & -5 + j5 \\ -5 + j5 & -5 + j5 & 10 - j10 \end{bmatrix}$$

Example 3: Find Y-Bus matrix of the following Circuits.

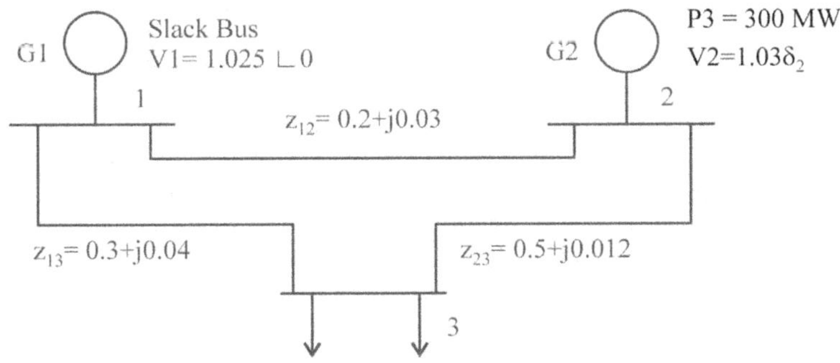

Fig.6.4 One line diagram of 3-Bus System.

Solution:

$$\begin{bmatrix} I_1 \\ I_2 \\ I_3 \end{bmatrix} = \begin{bmatrix} Y_{11} & Y_{12} & Y_{13} \\ Y_{21} & Y_{22} & Y_{23} \\ Y_{31} & Y_{32} & Y_{33} \end{bmatrix} \begin{bmatrix} v_1 \\ v_2 \\ v_3 \end{bmatrix}$$

$z_{12} = 0.2 + j0.03$

$z_{13} = 0.3 + j0.04$

$z_{23} = 0.5 + j0.012$

Admittance, $y_{12} = 1/Z_{12} = 1/(0.2+j0.03) = 4.889 - j0.733$ (rectangular form)

$y_{13} = 1/Z_{13} = 1/(0.3 + j0.04) = 3.275 - j0.4366$ (rectangular form)

$y_{23} = 1/Z_{23} = 1/(0.5+j0.012) = 1.998 - j0.0479$ (rectangular form)

For Y-bus matrix,

$Y_{11} = y_{12} + y_{13} = (4.889 - j0.733) + (3.275 - j0.4366) = 8.164 - j1.1696$

$= 8.247 \ ∟ -8.152$ (Polar form)

$Y_{12} = Y_{21} = -y_{12} = -(4.889 - j0.733) = 4.9436 \angle 171.473$ (Polar form)

$Y_{13} = Y_{31} = -y_{13} = -(3.275 - j0.4366) = 3.3039 \angle 172.406$ (Polar form)

$Y_{22} = y_{12} + y_{23} = (4.889 - j0.733) + (1.998 - j0.0479) = 6.887 - j0.7809$

$= 6.931 \angle -6.468$ (Polar form)

$Y_{23} = Y_{32} = -y_{23} = -(1.998 - j0.0479) = 1.998 \angle 178.626$ (Polar form)

$Y_{33} = y_{13} + y_{23} = (3.275 - j0.4366) + (1.998 - j0.0479) = 5.273 - j0.4845$

$= 5.295 \angle -5.2497$ (Polar form)

$$Y_{BUS} = \begin{bmatrix} 8.247\angle -8.152 & 4.9436\angle 171.473 & 3.3039\angle 172.406 \\ 4.9436\angle 171.473 & 6.931\angle -6.468 & 1.998\angle 178.626 \\ 3.3039\angle 172.406 & 1.998\angle 178.626 & 5.295\angle -5.2497 \end{bmatrix}$$

Example 4: Find Y-Bus matrix of the following Circuits

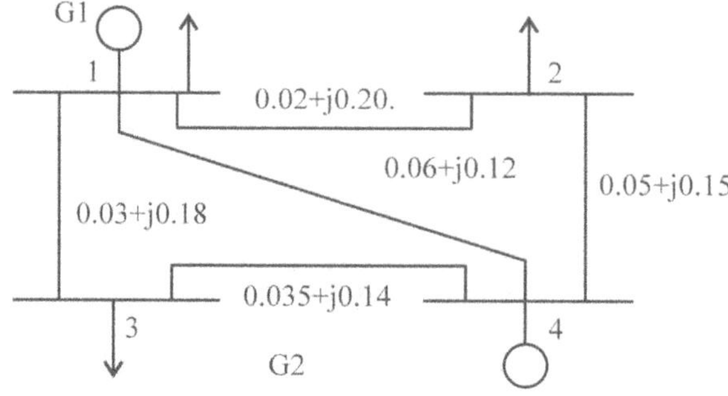

Fig.6.5 One line diagram of 4 Bus System.

Solution:

Let's take Base MVA = 100MVA

Given, $Z_{12} = 0.02 + j0.20$

$Z_{13} = 0.03 + j0.18$

$Z_{14} = 0.06 + j0.12$

$Z_{24} = 0.05 + j0.15$

$Z_{34} = 0.035 + j0.14$

Admittance, $y_{12} = 1/Z_{12} = 1/(0.02 + j0.20) = 0.495 - j4.950$ (rectangular form)

$y_{13} = 1/Z_{13} = 1/(0.03 + j0.18) = 0.900 - j5.4505$ (rectangular form)

$y_{14} = 1/Z_{14} = 1/(0.06 + j0.12) = 3.33 - j6.666$ (rectangular form)

$y_{24} = 1/Z_{24} = 1/(0.05 + j0.15) = 2 - j6$ (rectangular form)

$y_{34} = 1/Z_{34} = 1/(0.035 + j0.14) = 1.6806 - j6.722$ (rectangular form)

For Y-bus matrix,

$Y_{11} = y_{12} + y_{13} + y_{14} = (0.495 - j4.950) + (0.900 - j5.450) + (3.33 - j6.666)$

$= 4.725 - 17.066 = 17.658 \angle -74.524$ (Polar form)

$Y_{12} = Y_{21} = -y_{12} = - (0.495 - j4.950) = 4.974 \angle 95.710$ (Polar form)

$Y_{13} = Y_{31} = -y_{13} = - (0.900 - j5.405) = 5.479 \angle 99.453$ (Polar form)

$Y_{14} = Y_{41} = -y_{14} = - (3.33 - j6.666) = 7.451 \angle 116.544$ (Polar form)

$Y_{22} = y_{12} + y_{24} = (0.495 - j4.950) + (2 - j6) = 2.495 - j10.95$

$= 11.230 \angle -77.164$ (Polar form)

$Y_{24} = Y_{42} = -y_{24} = - (2 - j6) = 6.324 \angle 108.434$ (Polar form)

$Y_{33} = y_{13} + y_{34} = (0.900 - j5.405) + (1.6806 - j6.722) = 2.5806 - j12.127$

$= 12.398 \angle -77.986$ (Polar form)

$Y_{34} = Y_{43} = -y_{34} = - (1.6806 - j6.722) = 6.928 \angle 104.07$ (Polar form)

$Y_{44} = y_{14} + y_{24} + y_{34} = (3.33 - j6.666) + (2 - j6) + (1.6806 - j6.722)$

$= 7.0106 - j19.388 = 20.616 \angle -70.120$ (Polar form)

$$Y_{BUS} = \begin{bmatrix} 17.708\angle -74.480 & 4.974\angle 95.710 & 5.479\angle 99.453 & 7.451\angle 116.544 \\ 4.974\angle 95.710 & 11.230\angle -77.164 & 0 & 6.324\angle 108.434 \\ 5.479\angle 99.453 & 0 & 12.398\angle -77.986 & 6.928\angle 104.07 \\ 7.451\angle 116.544 & 6.324\angle 108.434 & 6.928\angle 104.037 & 20.616\angle -70.120 \end{bmatrix}$$

Example 5: Find the Y bus of the given Network.

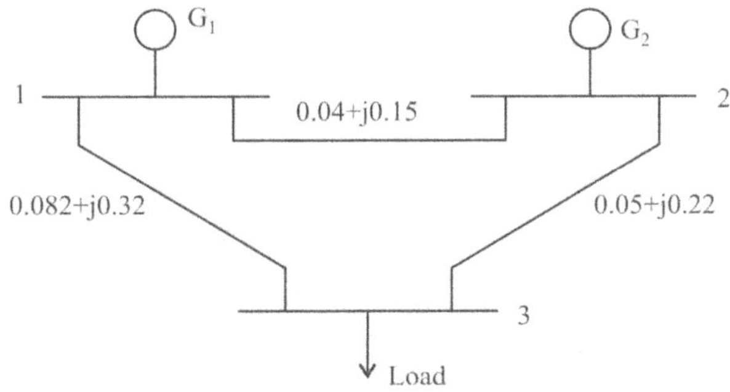

Fig.6.6 One line diagram of 3-Bus System.

Solution: $y_{12} = 1/(0.04 + j0.15) = (0.04 - j0.15)/0.0241 = 1.65 - j6.224$

$y_{13} = 1/(0.082 + j0.32) = (0.082 - j0.32)/0.097 = 0.751 - j2.93$

$y_{23} = 1/(0.05 + j0.22) = (0.05 - j0.22)/0.05 = 0.9823 - j4.3222$

Now,

$Y_{11} = y_{12} + y_{13} = 1.65 - j6.224 + 0.751 - j2.93 =$

$Y_{22} = y_{12} + y_{23} = 1.65 - j6.224 + 0.9823 - j4.3222$

$Y_{33} = y_{23} + y_{13} = 0.9823 - j4.3222 + 0.751 - j2.93$

$Y_{12} = - y_{21} = -y_{12} = -1.65 + j6.224$

$Y_{13} = -y_{31} = y_{13} = -0.751 + j2.93$

$Y_{23} = -y_{32} = -y_{23} = -0.9823 + j4.3222$

$$Y = \begin{bmatrix} Y_{11} & Y_{12} & Y_{13} \\ Y_{21} & Y_{22} & Y_{23} \\ Y_{31} & Y_{32} & Y_{33} \end{bmatrix}$$

$$Y = \begin{bmatrix} 2.401 - j9.154 & -1.65 + j6.224 & -0.751 + j2.93 \\ -1.65 + j6.224 & 2.6323 - j10.5462 & -0.9823 + j4.3222 \\ -0.751 + j2.93 & -0.9823 + j4.3222 & 1.7333 - j7.2522 \end{bmatrix}$$

6.5 The Algorithm for the Gauss-Seidel Iteration

We will now look at the algorithm for the Gauss-Seidel Iteration method for solving the system of equations $Ax = B$.

Take A be $n \times n$ matrix and B be an $n \times 1$ matrix.

Let system $Ax = B$ have solutions.

Take an initial assumption of the solution to the system

Also, take the number of iteration and the minimum error you want in the solution.

Now get the approximates, components wise,

$$x^0 = \begin{bmatrix} x_1^0 \\ x_2^0 \\ x_3^0 \\ x_4^0 \\ . \\ . \\ . \\ x_n^0 - 1 \\ x_n^0 \end{bmatrix} \qquad \qquad .....(1)$$

Now for closeness of the answer check the error present in the last iteration.

$$x^{k-x(k-1)} = \left| x_i^{(k)} - x_i^{(k-1)} \right| \text{error} \qquad\qquad .....(2)$$

If the above condition is true then the last iteration is assumed as the correct answer.

Gauss Seidel Acceleration

By introducing the acceleration factor to the solution obtained after each iteration, the speed of convergence can be expedited. The acceleration factor is a factor that enables solution correction between two subsequent iterations.

$$X_{acc}^{r+1} = X^{(r)} + \alpha[X^{(r+1)} - X^{(r)}]$$

During Gauss Seidel analysis, for any solution of variable X, the X_{acc}^{r+1} can be obtained by using the above expression where $X^{(r)}$ is the r the iterative value, α is acceleration factor and $X^{(r+1)}$ is the r the iterative value.

Example 6: Solve using Gauss Seidel method:

$$F(x) - x^3 - 5x^2 + 10x - 4 = 0$$

Solution: $x = -x^3/10 + x^2/2 + 2/5 = g(x)$

Initial guess

$$X^{(0)} = 2$$

1st iteration: $X^{(1)} = g(2) = -8/10 + 2 + 2/5 = 1.6$

2nd iteration: $X^{(2)} = g(1.6) = -(1.6)^3/10 + (1.6)^2/2 + 2/5 = 1.2704$

3rd iteration: $X^{(3)} = g(1.2704) = -(1.2704)^3/10 + (1.2704)^2/2 + 2/5 = 1.001926171$

4th iteration: $X^{(4)} = g(1.001926171) = -(1.001926171)^3/10 + (1.001926171)^2/2 + 2/5$

$$= 0.801349061$$

5th iteration: $X^{(5)} = g(0.801349061) = -(0.801349061)^3/10 + (0.801349061)^2/2 + 2/5$

$$= 0.669620702$$

6th iteration: $X^{(6)} = g(0.669620702) = -(0.669620702)^3/10 + (0.669620702)^2/2 + 2/5$

$$= 0.5941706934$$

7th iteration: $X^{(7)} = g(0.5941706934) = -(0.5941706934)^3/10 + (0.5941706934)^2/2 + 2/5$

$$= 0.5555428748$$

8th iteration: $X^{(8)} = g(0.5555428748) = -(0.5555428748)^3/10 + (0.5555428748)^2/2 + 2/5$

$$= 0.5371683406$$

9th iteration: $X^{(9)} = g(0.5371683406) = -(0.5371683406)^3/10 + (0.5371683406)^2/2 + 2/5$

$$= 0.5287749299$$

6.5.1 Importance of Gauss Seidel Method in Power Flow Analysis

1. It is one of the simplest iterative methods in power flow studies since early days of power system and is a modification of Gauss iteration method.

2. It can be conveniently use for load flow studies in small power systems.

3. It can give an initial value to Newton Raphson method for large power systems.

Power flow equations:

$$S_i^* = P_i - jQ_i \qquad\qquad .....(1)$$

$$= V_i^* \Sigma_{k=1}^{N} Y_{ik} V_k = V_i^* \left(\Sigma_{k=1.k\neq i}^{N} Y_{ik} V_k + Y_{ii} V_i \right) \qquad\qquad .....(2)$$

This equation can be written as:

$$V_i = \frac{1}{y_{ii}} \left[\frac{P_i - jQ_i}{V_i^*} - \Sigma_{k=1,k\neq i}^{N} Y_{ik} V_k \right] \qquad\qquad .....(3)$$

For i = 1, 2, 3,...., N

Using gauss-seidel method the final bus voltages after the iterations.

$$V_i^{p+1} = \frac{1}{Y_{11}} \left[\frac{P_i - jQ_i}{\left(V_i^p\right)^*} - \Sigma_{k=1,k\neq i}^{N} Y_{ik} V_k^p \right] \qquad\qquad .....(4)$$

As V_1 is known to be a prior one, iterations are to be performed for buses 2, 3, ..., N.

$$V_i = \frac{1}{Y_{ii}} \left[\frac{P_i - jQ_i}{V_i} - Y_{i1} V_1 - Y_{12} V_2 - Y_{i3} V_3 \cdots \cdots \cdots - Y_{in} V_n \right] \qquad\qquad .....(5)$$

Initially we assume the magnitude and angles at these (n-1) buses and update these values at each step of iteration until the desired value is reached.

6.5.2 Analysis

In power flow analysis, most of the equations are non-linear because of which its solving is very difficult. With large no. of equations some iterative method is very necessary to solve equations.

Gauss Seidel method is very approachable method. Almost accurate answer is approached but the number of iterations directly vary with the increase in the number of buses and that's why these method is limited to smaller number of Buses.

Example 7: Using Gauss-Seidel load flow method find bus voltages at the end of three iteration for the following system, shown in Fig.6.7. Line reactances are shown in Fig.6.7. Ignoring resistance and line charging. Assume initial voltage at all buses to $1.0 < 0^0$. Use 1.0 as acceleration factor.

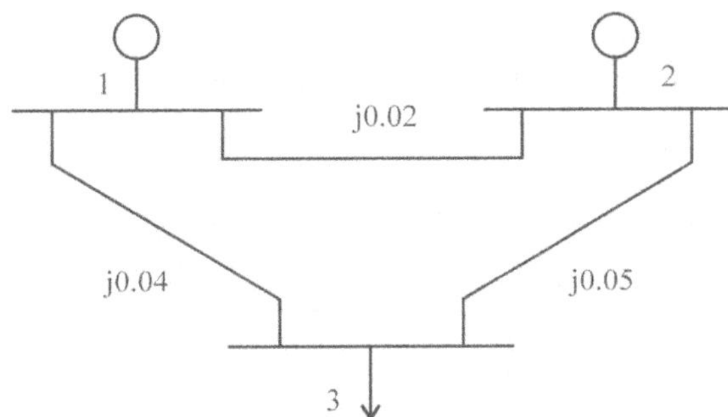

Fig. 6.7 Bus System.

Table 6.3 Bus Data.

Bus No.	P(pu)	Q(pu)	V(pu)
1	-	-	1.0
2	0.4	-	1.0
3	0.6	0.3	-

Solution: Admittance of each line,

$$y_{12} = \frac{1}{j0.02} = -j50, \; y_{13} = \frac{1}{j0.04} = -j25, \; y_{23} = \frac{1}{j0.05} = -j20$$

$Y_{11} = y_{12} + y_{13} = -j75, \; Y_{12} = -y_{12} = j50, \; Y_{21} = -y_{21} = j50, \; Y_{13} = -y_{13} = j25$

$Y_{22} = y_{21} + y_{23} = -j70, \; Y_{23} = -y_{23} = j20, \; Y_{31} = j25, \; Y_{32} = j20, \; Y_{33} = y_{31} + y_{32} = -j45$

The admittance matrix is given below

$$Y_{bus} = \begin{bmatrix} -j75 & j50 & j25 \\ j50 & -j70 & j20 \\ j25 & j20 & -j45 \end{bmatrix}$$

We have initial voltages as iteration number

$V_1^{(0)} = V_2^{(0)} = V_3^{(0)} = 1.0 \; \angle 0°$

$V_1^{(1)} = 1.0 \angle 0° = 1 + j0$ being specified (slack bus)

For the generator bus 2

$P_2 = 0.4$ pu.

$Q_2 = -jV_2(Y_{21}V_1 + Y_{22}V_2 + Y_{23}V_3) = jV_2(j50 - j70 + j20) = 0$

Now,

$V_2^{(1)} = 1/Y_{22} \, [(\, P_2 - jQ_2)/V_2^{(0)}*) - (Y_{21}V_1^{(0)} + Y_{23}V_3^{(0)})]$

$$= \frac{1}{-j70}\left[\frac{0.4}{1-j0} - j50 - j20\right]$$

$$= j0.014285(0.4 - j70)$$

$$= (1 + j0.00571428) \text{ p.u.}$$

$V_{2acc}{}^1 = V_2{}^{(0)} + \alpha[V_2{}^{(1)} - V_2{}^{(0)}] = (1.0 + j0) + (1 + j0.00571428 - 1 - j0) = 1 + j5.71428$

For the load bus 3

$V_3{}^{(1)} = 1/Y_{33}[((P_3 - jQ_3)/V_3{}^{(0)*}) - (Y_{31}V_1{}^{(0)} + Y_{32}V_2{}^{(0)})]$

$$= \frac{1}{-j45}\left[\frac{-0.6 + j0.3}{1 - j0} - j25 - j20\right]$$

$$= j0.02222[-0.6 - j44.7]$$

$$= (0.9933 - j0.01333)\text{pu}$$

$V_{3acc}{}^1 = V_3{}^{(0)} + \alpha[V_3{}^{(1)} - V_3{}^{(0)}] = (0.9933 - j0.01333)\text{pu}$

Now for 2nd iteration

For generator bus 2

$V_2{}^{(2)} = 1/Y_{22}[((P_2 - jQ_2)/V_2{}^{(1)*}) - (Y_{21}V_1{}^{(1)} + Y_{23}V_3{}^{(1)})]$

$$= \frac{1}{-j70}\left[\frac{0.4}{1 - j5.71428} - j50(1 + j0) - j20(1.0066 - j0.01333)\right]$$

$$= j0.014285 \times \left[\frac{0.4(1 + j5.71428)}{5.8011} - j50(1 + j0) - j20(1.0066 - j0.01333)\right]$$

$$= j0.014285 \times [0.06895 + j0.3940 - j50 - j20.132 - 0.2666]$$

$$= j0.014285 \times [-0.19765 - j69.738]$$

$$= (0.9962 - j2.85179)\text{p.u.}$$

$V_{2acc}{}^2 = V_2{}^{(1)} + \alpha[V_2{}^{(2)} - V_2{}^{(1)}] = (1 + j5.71428) + (0.9962 - j2.85179 - 1 - j5.71428)$

$$= (0.9962 - j2.85179)\text{p.u.}$$

For the load bus 3

$V_3{}^{(2)} = 1/Y_{33}[((P_3 - jQ_3)/ V_3{}^{(1)*}) - (Y_{31}V_1{}^{(1)} + Y_{32}V_2{}^{(1)})]$

$$= \frac{1}{-j45}\left[\frac{-0.6 - j0.3}{1.0066 + j0.01333} - j25(1 + j0) - j20(1 + j5.71428)\right]$$

$$= j0.02222 \times \left[\frac{(-0.6 - j0.3)(1.0066 - j0.01333)}{1.0134} - j25 - (-114.2856 + j20)\right]$$

$$= j0.02222 \times [0.59196 - j0.30580 - j25 + 114.2856 - 20j]$$

$$= j0.02222 \times [114.8775 - j45.3058]$$

$$= 2.5525 - j1.0066 \text{ p.u.}$$

$$V_{3acc}{}^2 = V_3{}^{(1)} + \alpha[V_3{}^{(2)} - V_3{}^{(1)}] = (0.83002 + j2.43148) \text{ p.u.}$$

Solution of the same problem by PSAT [1,2]

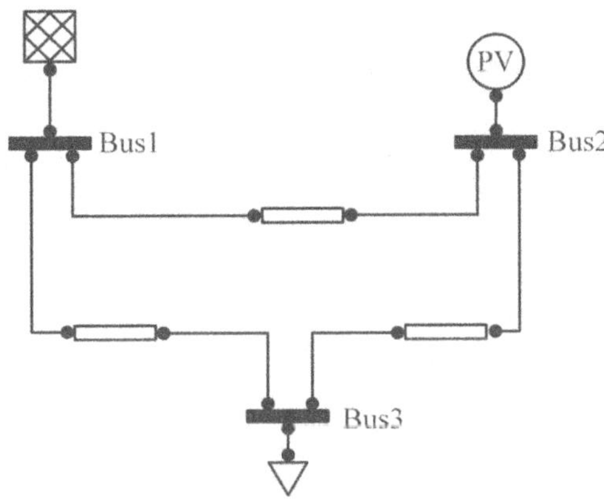

Fig.6.8

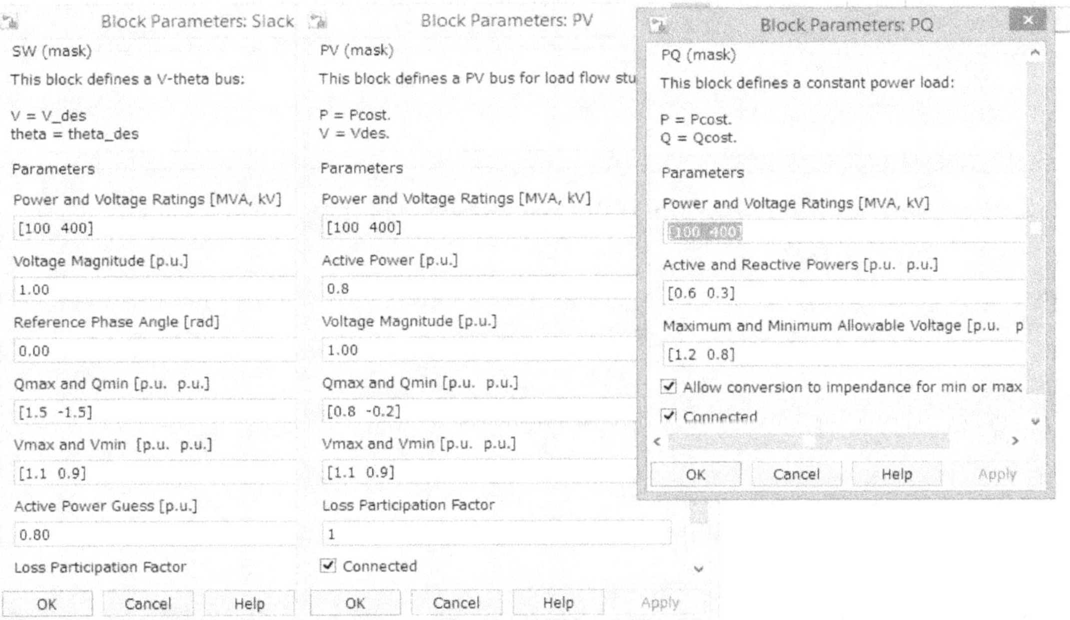

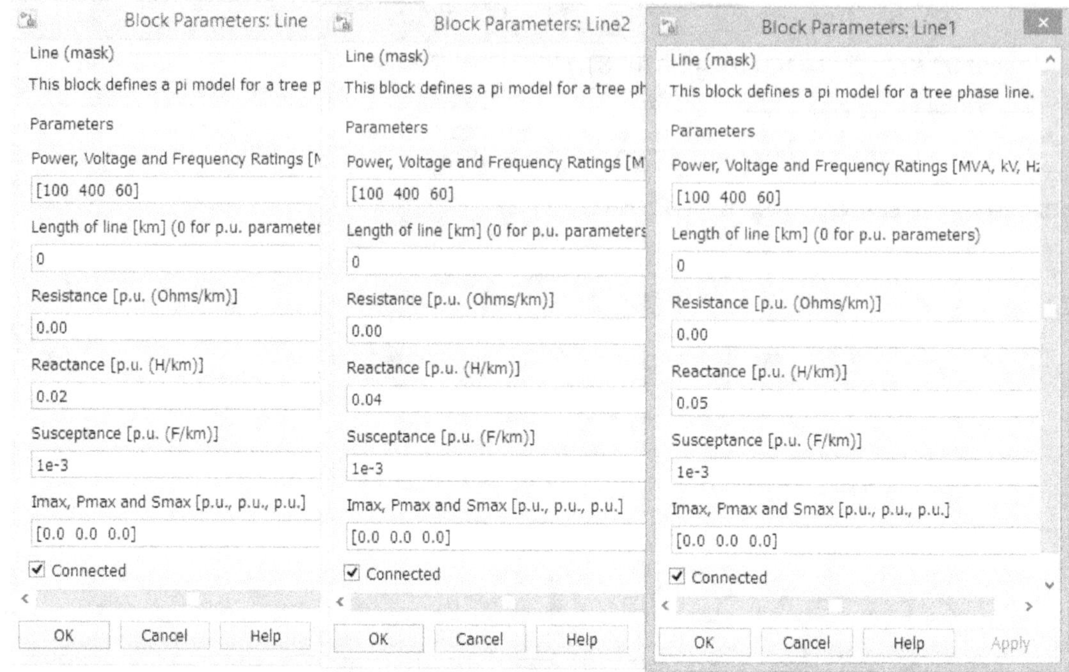

POWER FLOW REPORT

PSAT 2.1.9

Author: Federico Milano, (c) 2002-2013

e-mail: federico.milano@ucd.ie

website: faraday1.ucd.ie/psat.html

File: C:\Program Files\MATLAB\R2014a\bin\psat\tests\q6try.mdl

Date: 24-Jun-2019 17:18:25

NETWORK STATISTICS

Buses: 3

Lines: 3

Generators: 2

Loads: 1

SOLUTION STATISTICS

Number of Iterations: 3

Maximum P mismatch [p.u.] 0

Maximum Q mismatch [p.u.] 0

Power rate [MVA] 100

POWER FLOW RESULTS

Bus	V	phase	P gen	Q gen	P load	Q load
[p.u.]	[rad]	[p.u.]	[p.u.]	[p.u.]	[p.u.]	
Bus1	1	0	−0.2	0.17178	0	0
Bus2	1	0.00874	0.8	0.14002	0	0
Bus3	0.99321	−0.00954	0	0	0.6	0.3

LINE FLOWS

From Bus	To Bus	Line	P Flow	Q Flow	P Loss	Q Loss
		[p.u.]	[p.u.]	[p.u.]	[p.u.]	
Bus1	Bus2 1	-0.43691	0.00141	0		0.00282
Bus3	Bus2 2	-0.36309	-0.13205	0		0.00657
Bus1	Bus3 3	0.23691	0.17037	0		0.00242

LINE FLOWS

From Bus	To Bus	Line	P Flow	Q Flow	P Loss	Q Loss
			[p.u.]	[p.u.]	[p.u.]	[p.u.]
Bus2	Bus1	1	0.43691	0.00141	0	0.00282
Bus2	Bus3	2	0.36309	0.13861	0	0.00657
Bus3	Bus1	3	−0.23691	−0.16795	0	0.00242

GLOBAL SUMMARY REPORT

TOTAL GENERATION

REAL POWER [p.u.]   0.6

REACTIVE POWER [p.u.]   0.3118

TOTAL LOAD

REAL POWER [p.u.]   0.6

REACTIVE POWER [p.u.]   0.3

TOTAL LOSSES

REAL POWER [p.u.]   0

REACTIVE POWER [p.u.]   0.0118

Example 8: Using Gauss-Seidel load flow method, find bus voltages at the end of two iterations for the following 2 bus system. Line reactances are shown in Fig.(6.9). Ignore resistance and line charging. Assume initial voltage at all bus $1.0 \angle 0^0$. Use 1.0 as acceleration factor. The bus data has given in the table 6.4.

Table 6.4 Bus Data.

Bus no.	Specified P (pu)	Injections Q (pu)	Specified voltage (pu)
1	-	-	1.0
2	0.4	-	1.0
3	0.6	0.3	-

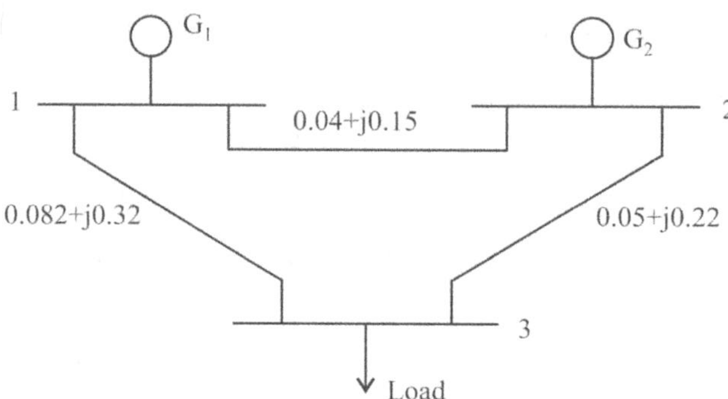

Fig.6.9 Bus System.

Solution:

$y_{12} = 1/(0.04 + j0.15) = (0.04 - j0.15)/0.0241 = 1.67 - j6.224$

$y_{13} = 1/(0.082 + j0.30) = (0.082 - j0.30)/0.097 = 0.845 - j3.09$

$y_{23} = 1/(0.05 + j0.22) = (0.05 - j0.22)/0.05 = 1 - j4.4$

$Y_{12} = -y_{12} = - -(1.67 - j6.224)$

$Y_{13} = -y_{13} = - -(0.845 - j3.09)$

$Y_{23} = -y_{23} = -(1 - j4.4)$

Now,

$Y_{11} = y_{12} + y_{13} = 2.515 - j9.314$

$Y_{22} = y_{12} + y_{23} = 2.67 - j10.624$

$Y_{33} = y_{13} + y_{23} = 1.845 - j7.49$

$$Y = \begin{bmatrix} 2.151 - j9.314 & -1.67 + j6.224 & -0.845 + j3.09 \\ -1.67 + j6.224 & 2.67 - j10.624 & -1 + j4.4 \\ -0.845 + j3.09 & -1 + j4.4 & 1.845 - j7.49 \end{bmatrix}$$

Now,

$V_1^{(0)} = V_2^{(0)} = V_3^{(0)} = 1.0 \angle 0^0$ as mentioned

So $V_1^{(1)} = 1.0 \angle 0^0$

$\quad\quad = 1 + j0$

For the generator bus 2

$P_2 = 0.4$

$Q_2 = 0$ (reference)

$V_2^{(1)} = 1/Y_{22}[(P_2 - jQ_2)/V_2^{(0)*} - (Y_{21}V_1^{(0)} + Y_{23}V_3^{(0)})]$

$\quad = 1/(2.67 - j10.624)\,[(0.4 - j0)/(1 - j0) + (1.67 - j6.224)(1.0 + j0) + (1 - j4.4)(1 + j0)]$

$\quad = (0.02225 + j0.0885)(3.07 - j10.624)$

$\quad = 1.009 + j0.039$

$V_2^{1}{}_{acc} = V_2^{(0)} + \alpha(V_2^{(1)} - V_2^{(0)})$

$\quad = (1 + j0) + 1(1.009 + j0.039 - 1 - j0)$

$\quad = 1.009 + j0.039$

$V_3^{(1)} = 1/Y_{33}[(P_3 - jQ_3)/V_3^{(0)*} - (Y_{31}V_1^{(0)} + Y_{32}V_2^{(0)})]$

$\quad = 1/(1.845 - j7.49)[(-0.6 + j0.3)/(1 - j0) + (0.845 - j3.09)(1 + j0) + (1 - j4.4)(1 + j0)]$

$\quad = (0.031 + j0.126)(1.245 - j7.19)$

$\quad = 0.9445 - j0.066$

$V_2^{(2)} = 1/Y_{22}[(P_2 - jQ_2)/V_2^{(1)*} - (Y_{21}V_1^{(1)} + Y_{23}V_3^{(1)})]$

$\quad = 1/(2.67 - j10.624)\,[(0.4 - j0)/(1.009 - j0.039)$

$\qquad + (1.67 - j6.224)\,(1.0 + j0) + (1 - j4.4)(0.9445 - j0.066)]$

$\quad = (0.02225 + j0.068)(3.301 - j10.432)$

$\quad = 0.997 + j0.060$

$V_3^{(2)} = 1/Y_{33}[(P_3 - jQ_3)/V_3^{(1)*} - (Y_{31}V_1^{(1)} + Y_{32}V_2^{(1)})]$

$\quad = 1/(1.845 - j7.49)\,[(-0.63 + j0.3)/(0.9445 + j0.066)$

$\qquad + (0.845 - j3.09)(1 + j0) + (1 - j4.4)(1.009 + j0.039)]$

$\quad = (0.031 + j0.126)(2.0256 - j7.4906)$

$\quad = 0.939 + j0.052$

$\quad = 0.5017 + j0.0230$

Solution of the same problem by PSAT [1,2]

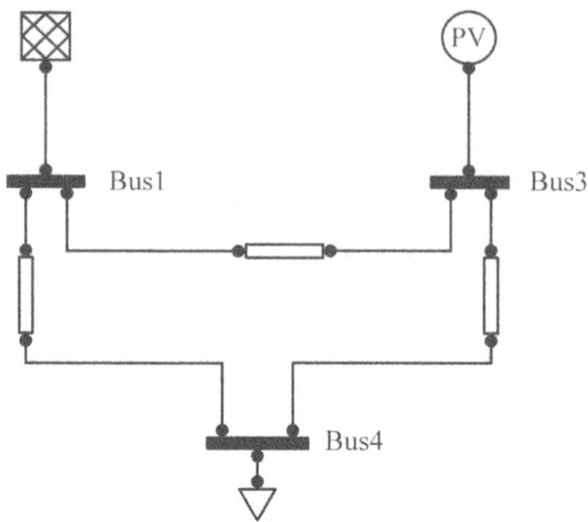

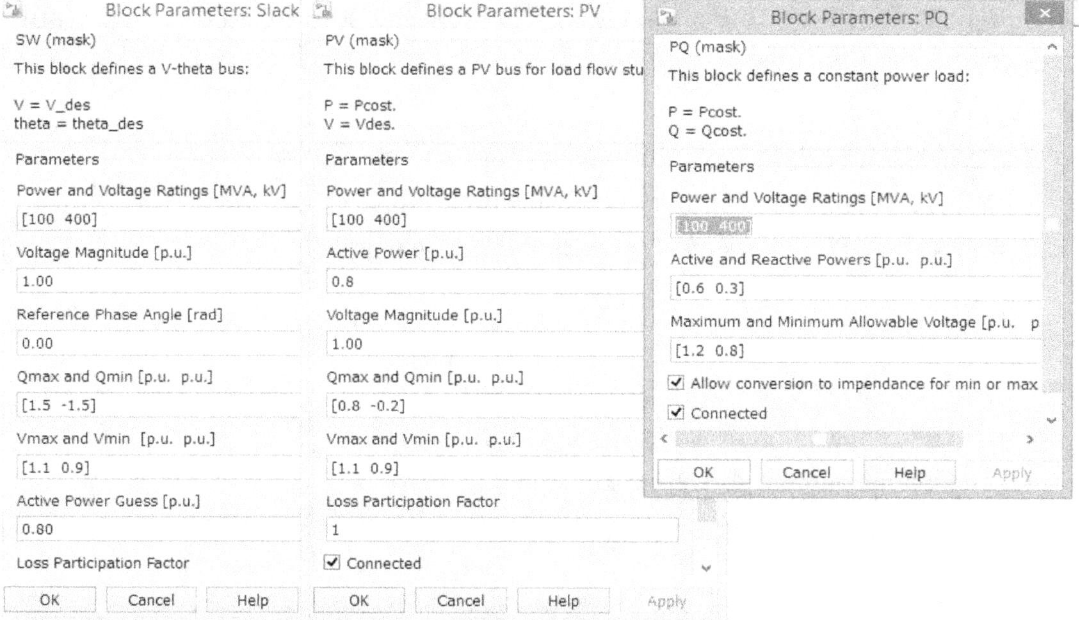

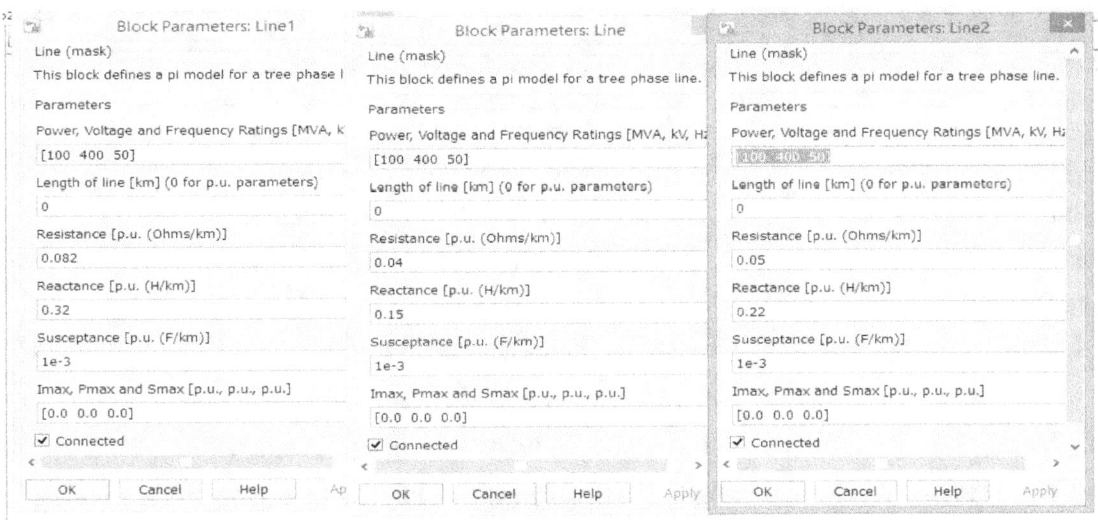

Author: Federico Milano, (c) 2002-2013

e-mail: federico.milano@ucd.ie

website: faraday1.ucd.ie/psat.html

File: C:\Program Files\MATLAB\R2014a\bin\psat\tests\prob2ps.mdl

Date: 21-Jun-2019 18:52:42

NETWORK STATISTICS

Buses: 3

Lines: 3

Generators: 2

Loads: 1

SOLUTION STATISTICS

Number of Iterations: 4

Maximum P mismatch [p.u.]: 0

Maximum Q mismatch [p.u.]: 0

Power rate [MVA]: 100

POWER FLOW RESULTS

Bus	V [p.u.]	Phase [rad]	P gen [p.u.]	Q gen [p.u.]	P load [p.u.]	Q load [p.u.]
Bus1	1	0	0.04309	0.39503	0	0
Bus3	1	0.04075	0.8	0.37833	0	0
Bus4	0.881	-0.07052	0	0	0.8	0.6

LINE FLOWS

From Bus	To Bus	Line	P Flow [p.u.]	Q Flow [p.u.]	P Loss [p.u.]	Q Loss [p.u.]
Bus1	Bus3	1	-0.25218	0.07228	0.00276	0.00933
Bus1	Bus4	2	0.29527	0.32275	0.01572	0.05661
Bus3	Bus4	3	0.54506	0.44128	0.02461	0.10741

LINE FLOWS

From Bus	To Bus	Line	P Flow [p.u.]	Q Flow [p.u.]	P Loss [p.u.]	Q Loss [p.u.]
Bus3	Bus1	1	0.25494	-0.06295	0.00276	0.00933
Bus4	Bus1	2	-0.27955	-0.26613	0.01572	0.05661
Bus4	Bus3	3	-0.52045	-0.33387	0.02461	0.10741

GLOBAL SUMMARY REPORT

TOTAL GENERATION

REAL POWER [p.u.] : 0.84309

REACTIVE POWER [p.u.]: 0.77336

TOTAL LOAD

REAL POWER [p.u.]: 0.8

REACTIVE POWER [p.u.] 0.6

TOTAL LOSSES

REAL POWER [p.u.]: 0.04309

REACTIVE POWER [p.u.] : 0.17336

Example 9: Find the voltage of bus using gauss seidel method.

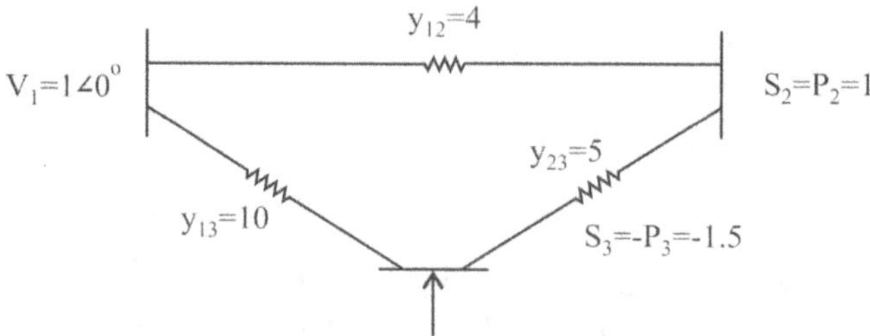

Fig.6.10 Bus System.

$$Y_{BUS} = \begin{bmatrix} Y_{11} & Y_{12} & Y_{13} \\ Y_{21} & Y_{22} & Y_{23} \\ Y_{31} & Y_{32} & Y_{33} \end{bmatrix}$$

$$= \begin{bmatrix} 14 & -4 & -10 \\ -4 & 9 & -5 \\ -10 & -5 & 15 \end{bmatrix}$$

Gauss Seidel Method

$$V_2^{i+1} = \frac{1}{Y_{22}}\left[\frac{S_2^*}{V_3^{i*}} - Y_{21}V_1 - Y_{23}V_3\right]$$

$$V_3^{i+1} = \frac{1}{Y_{33}}\left[\frac{S_3^*}{V_3^{i*}} - Y_{21}V_1 - Y_{23}V_3\right]$$

Table 6.5 Bus 2, Bus 3 Voltages at Different Iteration.

I	V_2	V_3
0	1	1
1	1.166	0.9
2	1.08777	0.9442
3	1.07118	0.923
4	1.06094	0.9154
5	1.0577	0.91107
6	1.0556	0.910

$V_2 = 1.05$ p.u. $V_3 = 0.91$ p.u

Power

$$S_2 = V_2(Y_{21}^{2*}V_1^* + Y_{22}^*V_2^* + Y_{23}^*V_3^*)$$

$$= 1.05(-4.1 + 9.1.05 - 5.0.91)$$

$$= 0.945$$

Example 10: The Fig.6.11 illustrates the single line diagram of a normal three bus power system. The generator is connected at bus number 1. The value of voltage at bus 1 is considered as 1.26 p.u. and a real power generation of 166.32 MW. A load of 307.92 MW and 132.24 Mvar is obtained from bus 2. Line impedances are expressed in per unit on a 100 MVA base. The line charging susceptances are avoided. Evaluate the Load flow using by the Gauss Seidel method.

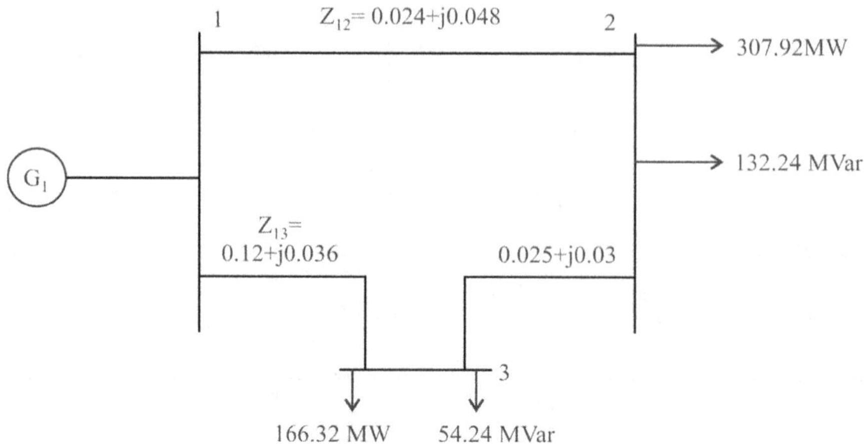

Fig.6.11 3-Bus Network.

Solution:

$$y_{12} = \frac{(0.024 - j0.048)10^3}{2.88}$$

$$= 8.3333 - j16.6666$$

$$= 18.62\angle{-63.43}$$

$$y_{13} = \frac{(0.012 - j0.036)(10)^3}{1.44}$$

$$= 8.33 - j25$$

$$= 26.35128\angle{-71.572}$$

$$y_{23} = \frac{(0.025 - j0.03)(10)^3}{1.525}$$

$$= 16.3934 - j19.6721$$

$$= 25.6\angle{-50.197}$$

$$Y_{11} = y_{12} + y_{13}$$

$$= (8.3333 - j16.6666) + (8.33 - j25)$$

$$= (15.9785 - j18.9601)$$

$$Y_{21} = -y_{12} = (-8.3333 + j\,16.6666)$$

$$Y_{31} = -y_{13} = (-8.33 - j25)$$

$$Y_{32} = -y_{23} = (-16.3934 + j\,19.6721)$$

$$Y_{22} = y_{21} + y_{23} = (8.3333 - j16.666) + (16.3934 - j19.6721)$$

$$= (24.7267 - j36.3387)$$

$$Y_{33} = y_{31} + y_{32} = (8.33 - j25) + (16.3934 - j19.6721)$$

$$= (24.0386 - J44.67)$$

$$Y_{bus} = \begin{bmatrix} (15.9785 - j18.9601) & (-8.3333 + j\,16.6666) & (-8.33 - j25) \\ (-8.3333 + j\,16.6666) & (24.7267 - j36.3387) & (-16.3934 + j\,19.6721) \\ (-8.33 - j25) & (-16.3934 + j\,19.6721) & (24.0386 - j21.9656) \end{bmatrix}$$

$$V_2^1 = \frac{\dfrac{P_2^{Sch} - jQ_2^{Sch}}{V_2^*} + Y_{21}V_1 + Y_{23}V_3^0}{Y_{23} + Y_{12}}$$

$$= \frac{\dfrac{-3.0792 + j1.3224}{1\angle 0^o} + (8.33 - j16.66)1.26 + (16.4 - j19.67)1\angle 0^o}{24.73 - j36.33}$$

$$= 1.1737 \angle\text{-}1.2929^o$$

$$= 1.0449 - j0.0557$$

$$V_2^2 = \frac{\dfrac{-3.0792 + j1.3224}{1.0449 - j0.0557} + (8.33 - j16.66)1.26 + (16.4 - j19.67)1\angle 0^o}{24.7267 - j36.3387}$$

$$= 1.238\angle\text{-}1.998^o$$

$$= 1.23732 - j0.043169$$

and

$$V_3^1 = \frac{\dfrac{P_3^{SCh} - jQ_3^{SCh}}{V_3^*} + Y_{13}V_1 + Y_{23}2}{Y_{13} + Y_{23}}$$

$$= \frac{\dfrac{-1.6632 + j0.5432}{1\angle 0^o} + (8.33 - j25)1.26\angle 0^o + \big((16.3934 + j19.6721)(1.0449 - j0.0557)\big)}{24.386 - j21.9656}$$

$$= \frac{32 - j53.8275}{24.386 - j21.9656}$$

$$= 1.107\angle\text{-}2.499^o$$

$$= 1.10679 - j0.048057$$

$$V_3^2 = \frac{\dfrac{-1.6632 + j0.5424}{1.0375 - j0.0426} + (8.33 - j25)(1.26) + (16.3934 + j19.6721)(1.0741 - j0.0796)}{24.0386 - j21.9656}$$

$$= \frac{62.45109428\angle\text{-}59.42540^o}{24.0386 - j21.9656}$$

$$= 0.994\angle\text{-}5.47^o$$

$$= 0.9903 - j0.094851$$

Solution of the same problem by PSAT [1,2]

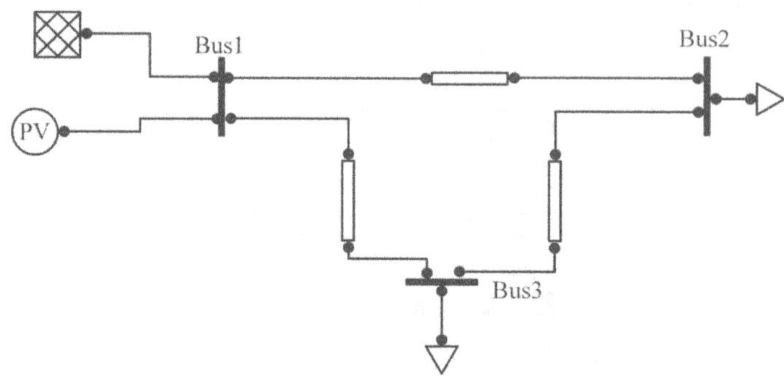

Fig.6.12

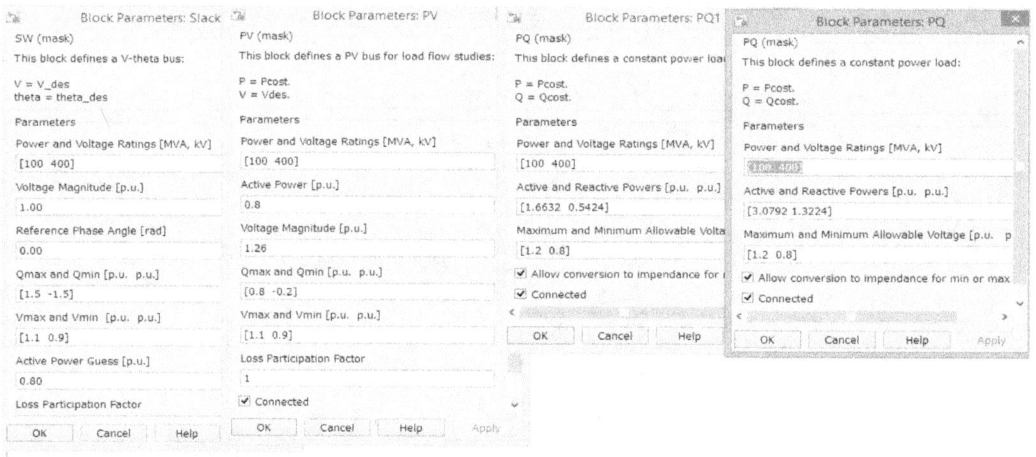

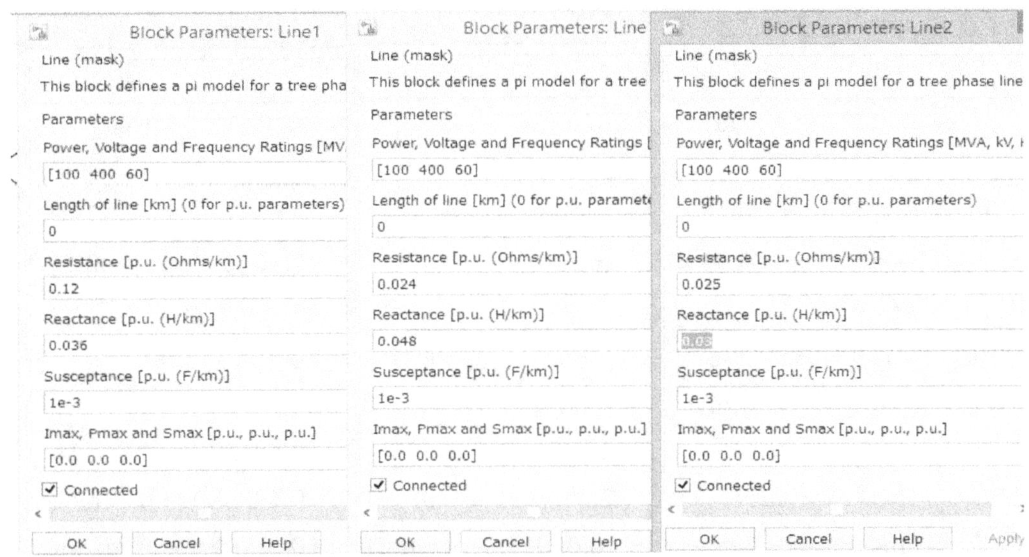

POWER FLOW REPORT

P S A T 2.1.9

Author: Federico Milano, (c) 2002-2013

e-mail: federico.milano@ucd.ie

website: faraday1.ucd.ie/psat.html

File: C:\Program Files\MATLAB\R2014a\bin\psat\tests\q9try.mdl

Date: 24-Jun-2019 16:14:19

NETWORK STATISTICS

Buses:	3
Lines:	3
Generators:	2
Loads:	2

SOLUTION STATISTICS

Number of Iterations:	5
Maximum P mismatch [p.u.]	0
Maximum Q mismatch [p.u.]	0
Power rate [MVA]	100

POWER FLOW RESULTS

Bus	V [p.u.]	phase [rad]	P gen [p.u.]	Q gen [p.u.]	P load [p.u.]	Q load [p.u.]
Bus1	1	0	5.4571	2.8952	0	0
Bus2	0.78477	-0.12216	0	0	2.9631	1.2725
Bus3	0.76058	-0.09447	0	0	1.5034	0.49027

LINE FLOWS

From Bus [p.u.]	To Bus [p.u.]	Line [p.u.]	P Flow [p.u.]	Q Flow	P Loss	Q Loss
Bus1	Bus2	1	3.4362	2.8873	0.48352	0.96624
Bus1	Bus3	2	2.0209	0.00791	0.49008	0.14623
Bus2	Bus3	3	-0.01035	0.64851	0.01709	0.01991

LINE FLOWS

From Bus [p.u.]	To Bus [p.u.]	Line [p.u.]	P Flow [p.u.]	Q Flow	P Loss	Q Loss
Bus2	Bus1	1	-2.9527	-1.921	0.48352	0.96624
Bus3	Bus1	2	-1.5308	0.13833	0.49008	0.14623
Bus3	Bus2	3	0.02744	-0.6286	0.01709	0.01991

GLOBAL SUMMARY REPORT

TOTAL GENERATION

REAL POWER [p.u.]   5.4571

REACTIVE POWER [p.u.]   2.8952

TOTAL LOAD

REAL POWER [p.u.]   4.4664

REACTIVE POWER [p.u.]   1.7628

TOTAL LOSSES

REAL POWER [p.u.]   0.99069

REACTIVE POWER [p.u.]   1.1324

Example 11: The one line diagram of a simple three bus power system with generation at bus 1. The magnitude of voltage at bus 1 is adjusted to 1.575 per unit. The scheduled loads at buses 2 and 3 are as marked on the diagram. Line impedances are marked in per unit on a 100 MVA base and the line charging susceptances are neglected. Using the Gauss Seidel method, determine the phasor values of the voltage at load buses 2 and 3.

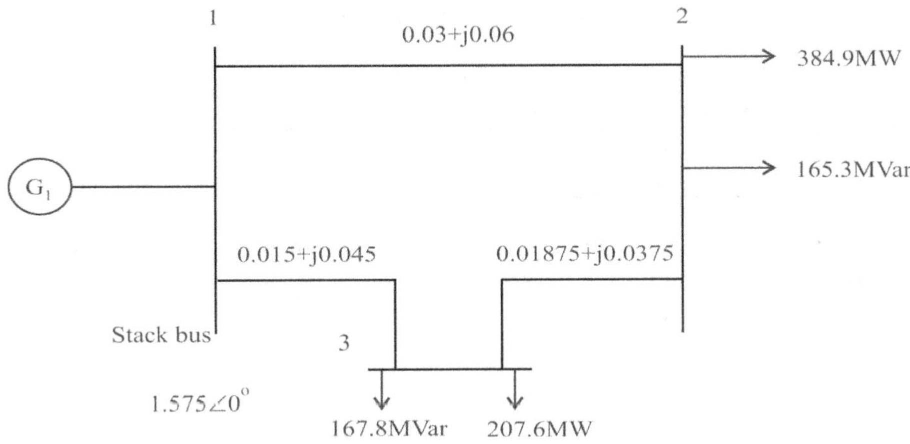

Fig. 6.13

Solution: Load impedances converted to admittances

$$y_{12} = \frac{1}{0.03+j0.06} = 6.666 - j13.333$$

$$y_{13} = \frac{1}{0.015+j0.045} = 6.666 - j20$$

$$y_{23} = \frac{1}{0.01875+j0.0375} = 10.666 - j21.3$$

$$Y_{21} = -y_{12} = -6.666 + j13.333$$

$$Y_{31} = -y_{13} = -6.666 + j20$$

$$Y_{32} = -y_{23} = -10.666 + j21.333$$

$$Y_{11} = y_{12} + y_{13} = (6.666 - j13.333) + (6.666 - j20)$$

$$= (13.333 - j33.333)$$

$$Y_{22} = y_{21} + y_{23} = (6.6666 - j13.3333) + (10.6666 - j21.3333)$$

$$= 17.3332 - j34.6666$$

$$Y_{33} = y_{31} + y_{32} = (6.6666 - j20) + (10.6666 - j21.3333)$$

$$= 17.3332 - j41.3333$$

At P-Q buses, complex loads are expressed in per units

$$S_2^{sch} = \frac{-(384.9+j165.3)}{100} = -3.849 - j1.653 \text{ pu}$$

$$S_3^{sch} = \frac{-(207.6+j167.8)}{100} = -2.076 - j1.678 \text{ pu}$$

Since bus 1 is taken as reference bus,

First iteration

$$V_2^1 = \frac{\dfrac{P_{sch}^2 - jQ_{sch}^2}{V_2^{0*}} + y_{21}V_1 + y_{23}V_3^0}{y_{21} + y_{23}}$$

$$= \frac{(-3.849+j1.653)/(1-j0)+(6.67-j13.33)(1.575+j0)+(10.67-j21.33)(1.0+j0)}{17.3332-j34.6666}$$

$$= \frac{-3.849+j1.653+10.50-j20.99+10.67-j21.33}{17.34-j34.66}$$

$$= \frac{17.321-j40.667}{17.34-j34.66} \times \frac{17.34+j34.66}{17.34+j34.66}$$

$$= \frac{300.346+j600.345-j705.165+1409.518}{1501.985}$$

$$= 1.138 - j0.0697$$

$$= 1.1407 \angle -0.0611^\circ$$

$$V_3^1 = \frac{\dfrac{P_{sch}^3 - jQ_{sch}^3}{V(0)_3^*} + y_{13}V_1 + y_{23}V_2^0}{y_{13}+y_{23}}$$

$$= \frac{(-2.076+j0.678)/(1-j0)+(6.67-j20)(1.575+j0)+(10.67-j21.33)(1-j0)}{17.34-j41.33}$$

$$= \frac{19.291-j55.73}{17.34-j41.33} \times \frac{17.34+j41.33}{17.34+j41.33}$$

$$= 0.2396 - j0.06296$$

$$= 0.24773 \angle -14.7228^\circ$$

Second iteration

$$V_2^2 = \frac{\dfrac{P_{sch}^2 - jQ_{sch}^2}{V_2^{1*}} + y_{21}V_1 + y_{23}V_3^1}{y_{21} + y_{23}}$$

$$= \frac{(-3.85+j1.65)/(1.138-j0.06)+22.722-j48.10}{17.34-j34.66}$$

$$= \frac{-3.449+j1.270+22.722-j48.10}{17.34-j34.66} \times \frac{17.34+j34.66}{17.34+j34.66}$$

$$= \frac{334.192+j667.999-j812.126+1623.316}{1501.99}$$

$$= 1.313 - j0.09$$

$$= 1.3211 \angle -3.921$$

$$V_3^2 = \frac{\frac{P_{sch}^3 - jQ_{sch}^3}{V_3^{1*}} + y_{13}V_1 + y_{23}V_2^2}{y_{13} + y_{23}}$$

$$= \frac{\frac{-2.076 + j0.678}{1.3131 - j0.0841} + (6.67 - j20)(1.575 + j0) + (10.67 - j21.33)(1.30327 - j0.095)}{17.34 - j41.33}$$

$$= \frac{1.607 - j0.4133 + j22.3845 - j60.3123}{17.34 - j41.33} \times \frac{17.34 + j41.33}{17.34 + j41.33}$$

$$= \frac{360.27 + j858.71 - j1038.648 + 2475.625}{2008.845}$$

$$= 1.3211 \angle - 3.921$$

Third iteration

$$V_2^3 = \frac{\frac{P_{sch}^2 - jQ_{sch}^2}{V_2^{2*}} + y_{13}v_1 + y_{23}v_3^2}{y_{21} + y_{23}}$$

$$= \frac{(-3.849 + j1.653)/(1.3330 - j0.115) + (6.67 - j13.33)(1.575 + j0) + (10.67 - j21.33)(1.411 - j0.089)}{17.34 - j34.66}$$

$$= \frac{-3.0297 + j1.0475 + 23.669 - j52.055}{17.34 - j34.66} \times \frac{17.34 + j34.66}{17.34 + j34.66}$$

$$= \frac{357.88 + j715.34 - j884.47 + 1767.91}{1501.99}$$

$$= 1.415 - j0.1126$$

$$V_3^3 = \frac{\frac{P_{sch}^2 - jQ_{sch}^3}{V_3^{2*}} + y_{13}v_1 + y_{23}v_2^3}{y_{13} + y_{23}}$$

$$V_3^3 = \frac{P3sch - jQ3sch/v3*(2) + y13v_1 + y23v2(3)}{y_{12} + y_{23}}$$

$$= \frac{\frac{(-2.076 + j0.678)}{(1.411 - j0.089)} + (6.67 - j20)(1.575 + j0) + (10.67 - j21.33)(1.415 - j0.1126)}{17.34 - j41.33}$$

$$= \frac{(-1.495 + j0.386 + 23.2045 - j62.889)}{17.34 - j41.33} \times \frac{17.34 + j41.33}{17.34 + j41.33}$$

$$= \frac{376.449 + j897.27 - j1083.808 + 2583.2655}{2008.8445}$$

$$= 1.473 - j0.0928$$

The final solution is

$V_2 = 1.415 - j0.1126$

$V_3 = 1.47 - j0.0928$

Solution of the same problem by PSAT [1,2]

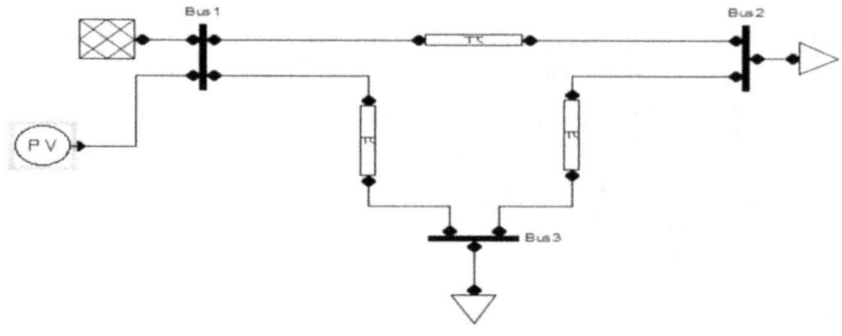

Fig.6.14

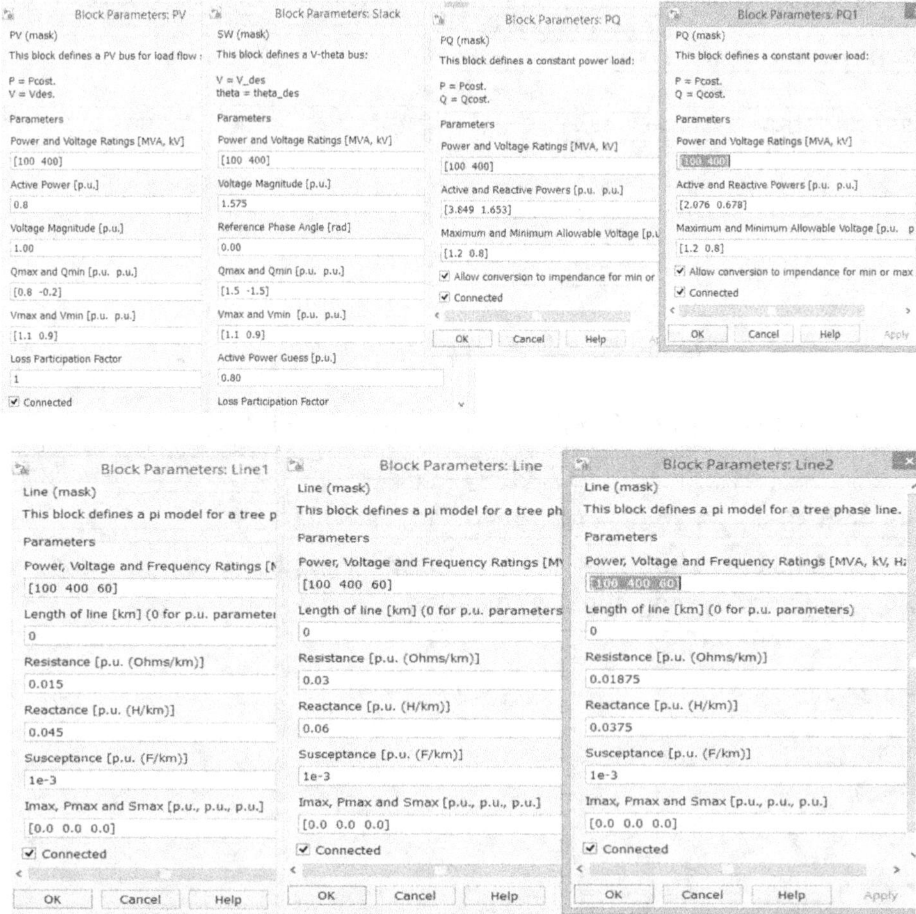

POWER FLOW REPORT

P S A T 2.1.9

Author: Federico Milano, (c) 2002-2013

e-mail: federico.milano@ucd.ie

website: faraday1.ucd.ie/psat.html

File: C:\Program Files\MATLAB\R2014a\bin\psat\tests\q10try.mdl

Date: 24-Jun-2019 11:49:57

NETWORK STATISTICS

Buses: 3

Lines: 3

Generators: 2

Loads: 2

SOLUTION STATISTICS

Number of Iterations: 6

Maximum P mismatch [p.u.]:   0

Maximum Q mismatch [p.u.]:   0

Power rate [MVA] :100

POWER FLOW RESULTS

Bus	V	phase	P gen	Q gen	P load	Q load
	[p.u.]	[rad]	[p.u.]	[p.u.]	[p.u.]	[p.u.]
Bus1	1.575	0	8.9711	4.4248	0	0
Bus2	1.423	-0.0902	0	0	5.4126	2.3245
Bus3	1.4647	-0.07425	00	3.093	1.0101	

LINE FLOWS

From Bus	To Bus	Line	P Flow	Q Flow	P Loss	Q Loss
			[p.u.]	[p.u.]	[p.u.]	
Bus1	Bus2	1	4.3484	1.9661	0.27548	0.54871
Bus1	Bus3	2	4.6227	2.4587	0.16581	0.49511
Bus2	Bus3	3	-1.3397	-0.90707	0.02422	0.04635

GLOBAL SUMMARY REPORT

TOTAL GENERATION

REAL POWER [p.u.]:	8.9711
REACTIVE POWER [p.u.]:	4.4248

TOTAL LOAD

REAL POWER [p.u.]:	8.5055
REACTIVE POWER [p.u.]:	3.3346

TOTAL LOSSES

REAL POWER [p.u.]:	0.46551
REACTIVE POWER [p.u.]:	1.0902

Example 12: Fig.6.15 shows the 1-line drawing of a plain three buses power system. The generators are connected at buses 1 and 3. The value of voltage at bus 1 is assumed to 1.575pu. Voltage value at bus 3 is set at 1.56pu with a real power generation of 300MW.A load consisting of 600MW and 375MVar is obtained from bus 2. Line impedances are represented in per unit on a 100MVA base and line charging susceptances are avoided. Evaluate the Load flow solution using Gauss Seidel method.

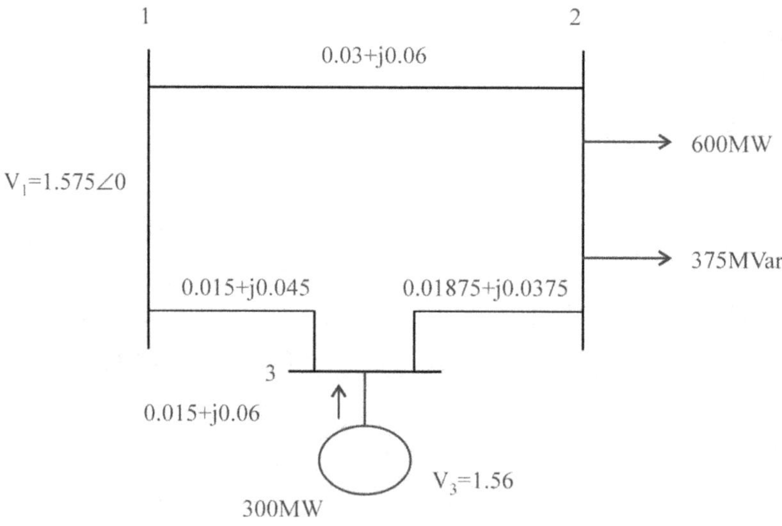

Fig.6.15

Solution: Admittances are

$$Y_{12} = \frac{1}{0.03+j0.06} = 6.67 - j13.33$$

$$Y_{13} = = \frac{1}{0.015 + j0.045} = 6.67 - j20$$

$$Y_{23} = \frac{1}{0.01875 + j0.0375} = 10.67 - j21.33$$

$$Y_{22} = y_{21} + y_{23} = 6.67 - j13.33 + 10.67 - j21.33$$

$$= 17.34 - j34.66$$

$$Y_{11} = y_{12} + y_{13} = 6.67 - j13.33 + 6.67 - j20 = 13.34 - j33.33$$

$$Y_{33} = y_{31} + y_{32} = 6.67 - j20 + 10.667 - j21.33 = 17.34 - j41.33$$

Load and generation expressed in per units

$$S_2^{sch} = -\frac{(600 + j375)}{100} = (-6 - j3.75) \text{ pu}$$

$$P_3^{sch} = \frac{300}{100} = 3 \, pu$$

Take bus 1 as reference bus.

$$V_2^{(0)} = 1.0 + j0 \text{ and } V_3^{(0)} = 1.56 + j0$$

$$V_2^1 = \frac{\frac{P_{sch}^2 - jQ_{sch}^2}{V_2^{0*}} + y_{12}V_1 + y_{23}V_3^0}{y_{12} + y_{23}}$$

$$= \frac{(-6 + j3.75)/(1.0 - j0) + (6.67 - j13.33)(1.575) + (10.67 - j21.33)(1.56)}{17.34 - j34.66}$$

$$= \frac{(21.16 - j50.51)}{17.34 - j34.66} \times \frac{17.34 + j34.66}{17.34 + j34.66}$$

$$= \frac{374.54 + j748.65 - j878.84 + 1750.67}{1501.99}$$

$$= 1.4111 - j0.0946$$

$$= 1.42 \angle - 0.0607$$

Bus 3 is regulated bus, where voltage magnitude and real power are specified. The first reactive power is computed

$$Q_3^{(1)} = -\{V_3^{*(0)} [V_3^{(0)}(y_{13} + y_{23}) - y_{13}V_1 - y_{23}V_2^{(1)}]\}$$

$$= -\{(1.56 - j0)[(1.56)(17.34 - j41.33) - (6.67 - j20)(1.575) - (10.67 - j21.33)(1.4099 - j0.0953)]\}$$

$$= -\{(1.56) (3.4565 - j1.8667)\}$$

$$= -5.39214 + j2.9121$$

$$= 2.9121$$

$$V_3^1 = \frac{\frac{P_{sch}^3 - jQ_{sch}^3}{V(0)_3^*} + y_{13}V_1 + y_{23}V_2^1}{y_{13} + y_{23}}$$

$$= \frac{(3 - j2.96)/1.56 + (6.67 - j20)(1.575) + (10.67 - j21.33(1.414 - j0.085))}{17.34 - j41.33}$$

$$= \frac{25.467 - j64.475}{17.34 - j41.33} \times \frac{17.34 + j41.33}{17.34 + j41.33}$$

$$= 1.0787 + j0.03267$$

Imaginary part, $f_3^{(1)} = 0.0214$

Real part, $e_3^{(1)} = 1.5460$

$$V_3^{(1)} = 1.5460 - j0.0331$$

$$= 1.5464 \angle -0.0214$$

Second iteration

$$V_2^2 = \frac{\frac{P_{Sch}^2 - jQ_{Sch}^2}{V_2^{1*}} + y_{12}V_1 + y_{23}V_3^1}{y_{12} + y_{23}}$$

$$= \frac{-6 + j3.75/(1.414 + j0.085) + (6.67 - j13.33)(1.575) + (10.67 - j21.33)(1.55 - j0.244)}{17.34 - j34.66}$$

$$= \frac{-4.06 + j2.89 + 10.505 - j20.99 + 21.743 - j30.45}{17.34 - j34.66}$$

$$= \frac{28.188 - j48.55}{17.34 - j34.66} \times \frac{17.34 + j34.66}{17.34 + j34.66}$$

$$= 1.2601 + j0.2087$$

$$= 1.304 \angle -0.164$$

$$Q_3^{(2)} = -\{V_3^{*(1)} [V_3^{(0)} (y_{13} + y_{23}) - y_{13}V_1 - y_{23} V_2^{(1)}]$$

$$= -\{(1.55 - j0.2442)[(1.55 - j0.2442)(17.33 - j41.33)$$

$$- (6.67 - j20)(1.575) - (10.67 - j21.33)(1.445 - j0.089)]\}$$

$$= -\{(1.55 + j0.148)(-2.036 - j2.59)\}$$

$$= 3.655 - j2.1464$$

$$= 2.1464$$

$$V_{c3}^2 = \frac{\frac{P_{Sch}^3 - jQ_{Sch}^3}{V_3^{1*}} + y_{13}V_1 + y_{23}V_2^2}{y_{13} + y_{23}}$$

$$= \frac{(3 + j0.160)/(1.55 + j0.2442) + (6.67 - j20)(1.575) + (10.67 - j21.33)(1.445 + j0.089)}{17.34 - j41.33}$$

$$= \frac{1.9044 - j0.1968 + 10.505 - j31.5 + 17.316 - j29.872}{17.34 - j41.33}$$

$$= \frac{29.7254 - j61.5685}{17.34 - j41.33} \times \frac{17.34 + j41.33}{17.34 + j41.33}$$

$$= 1.506 - j0.0195$$

$$= 1.506 \angle -0.0129$$

Imaginary part, $f_3^{(2)} = -0.0195$

Real part, $e_3^{(2)} = 1.506$

$V_3^{(2)} = 1.506 - j0.0195$

Third iteration

$$V_2^3 = \frac{\frac{P_{sch}^2 - jQ_{sch}^2}{V_2^{2*}} + y_{21}V_1 + y_{23}V_3^2}{y_{21} + y_{23}}$$

$$= \frac{(-6+j3.75)/(1.45-j0.15)+(6.67-j13.33)(1.575)+(10.67-j21.33)(1.558-j0.075)}{17.34-j34.66}$$

$$= \frac{-3.83+j2.89+10.505-j20.99+16.62-j0.8-j33.23-1.599}{17.34-j34.66}$$

$$= \frac{21.04-j49.8576}{17.34-j34.66} \times \frac{17.34+j34.66}{17.34+j34.66}$$

$$= 1.75 - j0.0896$$

$$= 1.75 \angle - 0.051$$

$$Q_3^{(3)} = -\{V_3^{*(2)} [V_3^{(2)} (y_{13} + y_{23}) - y_{13}V_1 - y_{23}V_2^{(3)}]$$

$$= -\{(1.0787 + j0.0327) [(1.558 - j0.075)(17.33 - j41.33)$$

$$- (6.67 - j20)(1.575) - (10.67 - j21.33)(1.45 - j0.1)]\}$$

$$= -\{(1.079 + j0.0327)(-0.983 + j5.3112)\}$$

$$= -\{-0.88631 + j5.762$$

$$= 5.762$$

$$V_{c3}^3 = \frac{\frac{P_{sch}^3 - jQ_{sch}^3}{V_3^{2*}} + y_{13}V_1 + y_{23}V_2^3}{y_{13} + y_{23}}$$

$$= \frac{(3-j3.41)/(1.558+j0.075)+(6.67-j20)(1.575)+(10.67-j21.33)(1.45-j0.1)}{17.34-j41.33}$$

$$= \frac{1.816-j2.276+10.505-j31.5+15.47-j1.067-j30.93-2.133}{17.34-j41.33}$$

$$= \frac{29.308-j78.4055}{17.34-j41.33} \times \frac{17.34+j41.33}{17.34+j41.33}$$

$$= 1.86 - j0.075$$

$$= 1.86 \angle - 0.0403$$

Imaginary part $f_3^{(3)} = -0.075$

Real part, $e_3^{(2)} = 1.86$

$V_3^{(2)} = 1.559 - j0.039$

Thus the Gauss Seidel method is solved successfully.

Solution of the same problem by PSAT [1,2]

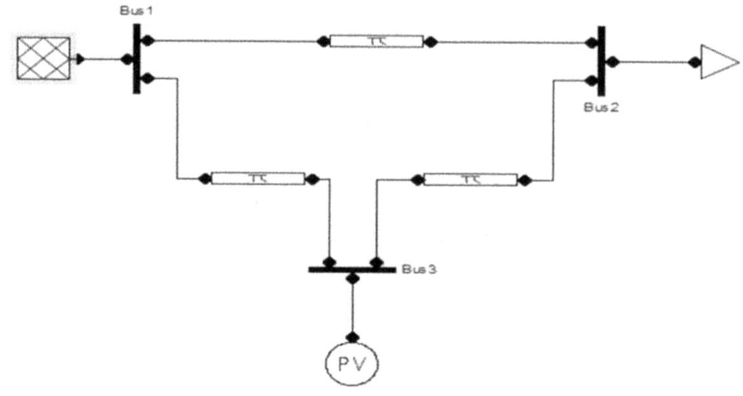

Fig. 6.16

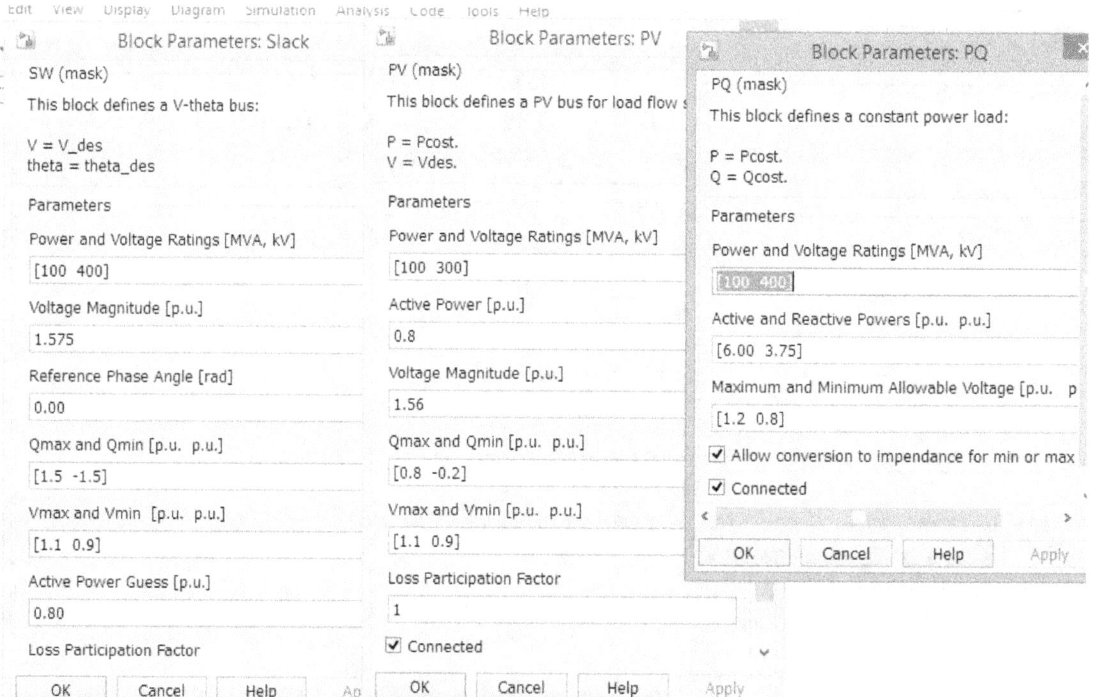

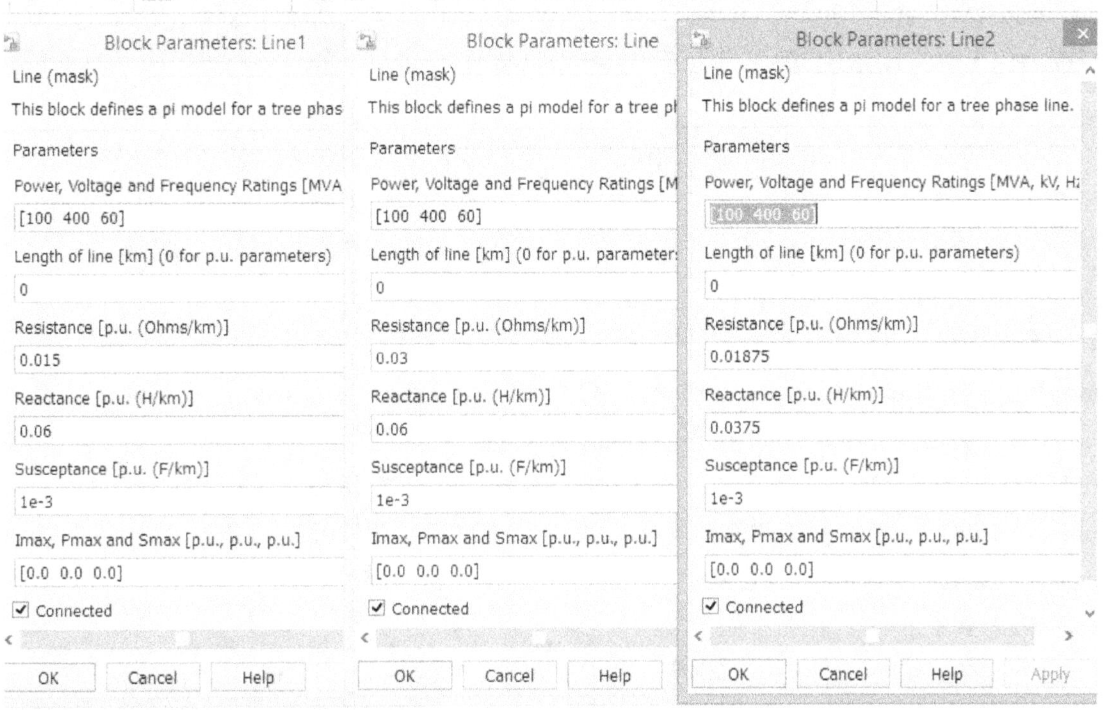

POWER FLOW REPORT

P S A T 2.1.9

Author: Federico Milano, (c) 2002-2013

e-mail: federico.milano@ucd.ie

website: faraday1.ucd.ie/psat.html

File: C:\Program Files\MATLAB\R2014a\bin\psat\tests\q12try.mdl

Date: 24-Jun-2019 20:43:03

NETWORK STATISTICS

Buses: 3

Lines: 3

Generators: 2

Loads: 1

SOLUTION STATISTICS

Number of Iterations: 6

Maximum P mismatch [p.u.]   0

Maximum Q mismatch [p.u.]   0

Power rate [MVA]   100

POWER FLOW RESULTS

Bus	V	phase	P gen	Q gen	P load	Q load
	[p.u.]	[rad]	[p.u.]	[p.u.]	[p.u.]	[p.u.]
Bus1	1.575	0	7.6833	3.0202	0	0
Bus2	1.361	-0.10589	0	0	7.7177	4.8236
Bus3	1.56	-0.0578	0.8	3.395	0	0

LINE FLOWS

From Bus	To Bus	Line	P Flow	Q Flow	P Loss	Q Loss
			[p.u.]	[p.u.]	[p.u.]	[p.u.]
Bus1	Bus2	1	5.348	3.1431	0.46547	0.92876
Bus1	Bus3	2	2.3353	-0.12293	0.03307	0.12981
Bus3	Bus2	3	3.1022	3.1423	0.26712	0.53304

LINE FLOWS

From Bus	To Bus	Line	P Flow	Q Flow	P Loss	Q Loss
			[p.u.]	[p.u.]	[p.u.]	[p.u.]
Bus2	Bus1	1	-4.8826	-2.2143	0.46547	0.92876
Bus3	Bus1	2	-2.3022	0.25274	0.03307	0.12981
Bus2	Bus3	3	-2.8351	-2.6092	0.26712	0.53304

GLOBAL SUMMARY REPORT

TOTAL GENERATION

REAL POWER [p.u.]:   8.4833

REACTIVE POWER [p.u.]:   6.4152

TOTAL LOAD

REAL POWER [p.u.]:   7.7177

REACTIVE POWER [p.u.]:   4.8236

TOTAL LOSSES

REAL POWER [p.u.]:   0.76566

REACTIVE POWER [p.u.]:   1.5916

Example 13: The one line diagram is given below of simple three buses. The generators are connected at buses 1 and 3.The value of voltage at buses 1 and 3 are set to 1.025pu and 1.03pu respectively. Bus 3 has active   power generation of 300MW. A load consisting of 400MW and 200MVAR is obtained from bus 2. Obtain the Load flow using by Gauss Seidel method.

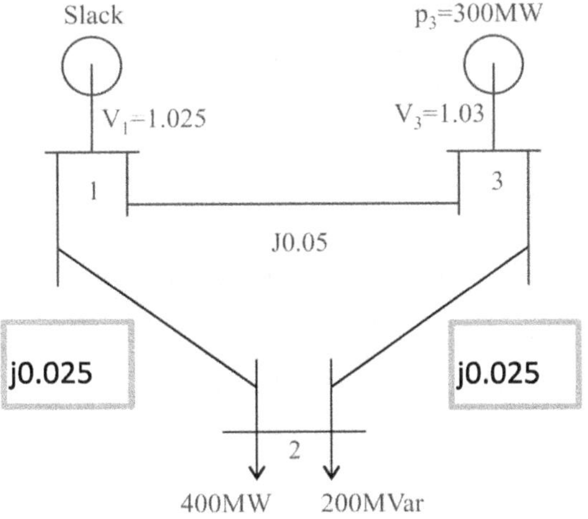

Fig. 6.17

Solution: Admittances are

$$Y_{12} = \frac{1}{0+j0.025} = 0 - j40$$

$$Y_{13} = \frac{1}{0+j0.05} = 0 - j20$$

$$Y_{23} = \frac{1}{0+j0.025} = 0 - j40$$

The load generators are expressed in per units

$$S_2^{sch} = \frac{-(400+j200)}{100} = -4 - j2 \text{ p.u.}$$

$$P_3^{sch} = 300 \text{ p.u.}$$

Bus 1 is taken as reference bus

$$V_1 = 1.025 + j0 \qquad V_2^{(0)} = 1.0 + j0 \qquad V_3^{(0)} = 1.03 + j0$$

First iteration

$$V_2^{(1)} = \frac{\dfrac{P_{sch}^2 - jQ_{sch}^2}{V_2^{*(0)}} + y_{12}V_1 + y_{23}V_3^{(0)}}{y_{12} + y_{23}}$$

$$= \frac{\dfrac{(-4+j2)}{(1.0-j0)} + (-j40)(1.025-j0)(-j40)(1.03+j0)}{0-j80}$$

$$= \frac{-4-j77.945}{0-j80} \times \frac{0+j80}{0+j80}$$

$$= -j0.0125(-4 - j77.945)$$

$$= 0.974 - j0.05$$

$$Q_3^{(1)} = -\{V_3^{*(0)}\,[V_3^{(0)}\,(y_{13} + y_{23}) - y_{13}V_1 - y_{23}V_2^{(0)}$$

$$= -\{(1.03)\,[1.03(0 - j60) - (-j20)(1.025 + j0) - (-j40)(0.974 - j0.05\ )]\}$$

$$= -\{(1.03)(2 - j2.34)\}$$

$$= -2.06 - j2.41$$

$$= 2.41$$

$$V_{c3}^{(1)} = \frac{\dfrac{P_3^{sch} = jQ_{sch}^2}{V_3^{*(0)}} + y_{13}V_1 + y_{23}V_2^{(1)}}{y_{13} + y_{23}}$$

$$= \frac{(3 - j1.339)/(1.03 - j0) + (-j20)(1.025 + j0) + (-j40)(0.49 - j0.05)}{0 - j60}$$

$$= \frac{2.912 - j1.3 - j20.5 - 2 - j19.6}{0 - j60} \times \frac{0 + j60}{0 + j60}$$

$$= \frac{0.912 - j414}{0 - j60}$$

$$= 0.146 - j1.032$$

Imaginary part $f_3^{(1)} = 1.032$

Real part, $e_3^{(1)} = 0.146$

$$V_3^{(1)} = 0.146 - j1.032$$

Second iteration

$$V_2^{(2)} = \frac{\dfrac{P_{sch}^2 - jQ_{sch}^2}{V_2^{*(1)}} + y_{12}V_1 + y_{23}V_3^{(1)}}{y_{12} + y_{23}}$$

$$= \frac{(-4 + j2)/(1.0025) + (-j40)(1.025 + j0) + (-j40)(0.146 - j1.032)}{0 - j80}$$

$$= \frac{(-3.99 + j1.9950) + (-j41) + (0.608 - j27.6)}{0 - j80} \times \frac{0 + j80}{0 + j80}$$

$$= \frac{-3.382 - j66.605}{0 - j80}$$

$$= 0.0823 - j0.0041$$

$$Q_3^{(2)} = -\{V_3^{*(1)}\,[V_3^{(1)}\,(y_{13} + y_{23}) - y_{13}V_1 - y_{23}V_2^{(2)}$$

$$= -\{(1.03 - j0.0152)[(1.03 + j0.0152)(0 - j60) - (-j20)(1.025 + j0)$$

$$- (-j40)(1.00 - j0.040)]\}$$

$$= -\{(1.03 - j0.0152)(2.512 - j1.3)\}$$

$$= -(9.295 + j42.783)$$

$$= 42.783$$

$$V_{c3}^{(2)} = \frac{\dfrac{P_{sch}^3 - jQ_{sch}^3}{V_3^{*(1)}} + y_{13}V_1 + y_{23}V_2^{(1)}}{y_{13} + y_{23}}$$

$$= \frac{(3 - j1.377)/(1.03 - j0.0152) + (-j20)(1.025 + j0) + (-j40)(0.974 - j0.05)}{0 - j60}$$

$$= \frac{(2.93 - j1.37 - 20.5 - j40 - 1.636)}{0 - j60} \times \frac{0 + j60}{0 + j60}$$

$$= \frac{2482.2 + j77.4}{3600}$$

$$= 0.7007 + j0.952$$

Imaginary part, $f_3^{(2)} = 0.952$

Real part, $e_3^{(2)} = 0.7007$

$$V_{c3}^{(2)} = 0.7007 + j0.952$$

Third iteration

$$V_2^{(3)} = \frac{\dfrac{P_{sch}^2 - jQ_{sch}^2}{V_2^{*(2)}} + y_{12}V_1 + y_{23}V_3^{(2)}}{y_{12} + y_{23}}$$

$$= \frac{(-4 + j2)/(1.00 - j0.0409) + (-j40)(1.025 + j0) + (-j40)(0.7007 + j0.952)}{0 - j80}$$

$$= \frac{(-32.92 - j70.1924)}{0 - j80} \times \frac{0 + j80}{0 + j80}$$

$$= -j0.0125(-32.92 - j70.1924)$$

$$= 0.877 - j0.412$$

$$Q_3^{(3)} = -\{V_3^{*(2)} [V_3^{(2)}(y_{13} + y_{23}) - y_{13}V_1 - y_{23}V_2^{(3)}]$$

$$= -\{(0.7007 - j0.9522)[(1.03 + j0.0152)(0 - j60) - (-j20)(1.025 + j0)$$

$$\qquad - (-j40)(0.877 - j0.412)]\}$$

$$= -\{(0.7007 - j0.9522)[-48.7 - j27.502]\}$$

$$= 60.3 - j27.9$$

$$= 27.9$$

$$V_{c3}^{(3)} = \frac{\dfrac{P_{sch}^3 - jQ_{sch}^3}{V_3^{*(2)}} + y_{13}V_1 + y_{23}V_2^{(2)}}{y_{13} + y_{23}}$$

$$= \frac{(3 - j1.3784)/(1.03 - j0.0215) + (-j20)(1.025 = j0) + (-j40)(1.0005 - j0.038)}{0 - j60}$$

$$= \frac{(2.939 - j1.2756 - j60.52 - 1.52)}{0 - j60} \times \frac{0 + j60}{0 + j60}$$

$$= j0.067(-9.50415 - j54.72)$$

$= 0.914\text{-}j0.1587$

Imaginary part, $f_3^{(3)} = 0.1587$

Real part, $e_3^{(3)} = 0.914$

$V_3^{(3)} = 0.914\text{-}j0.1587$

Solution of the same problem by PSAT [1,2]

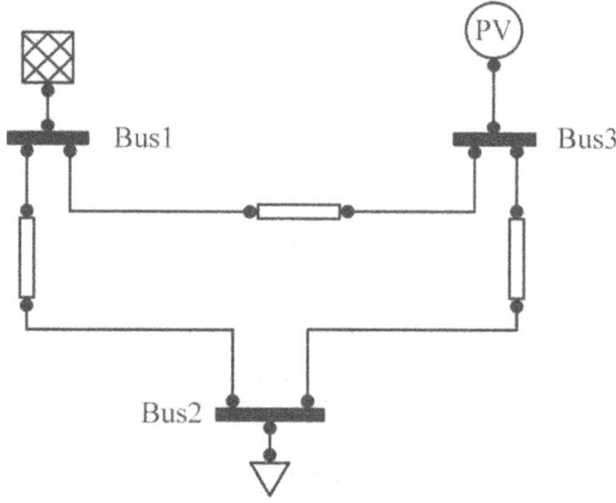

Fig. 6.18

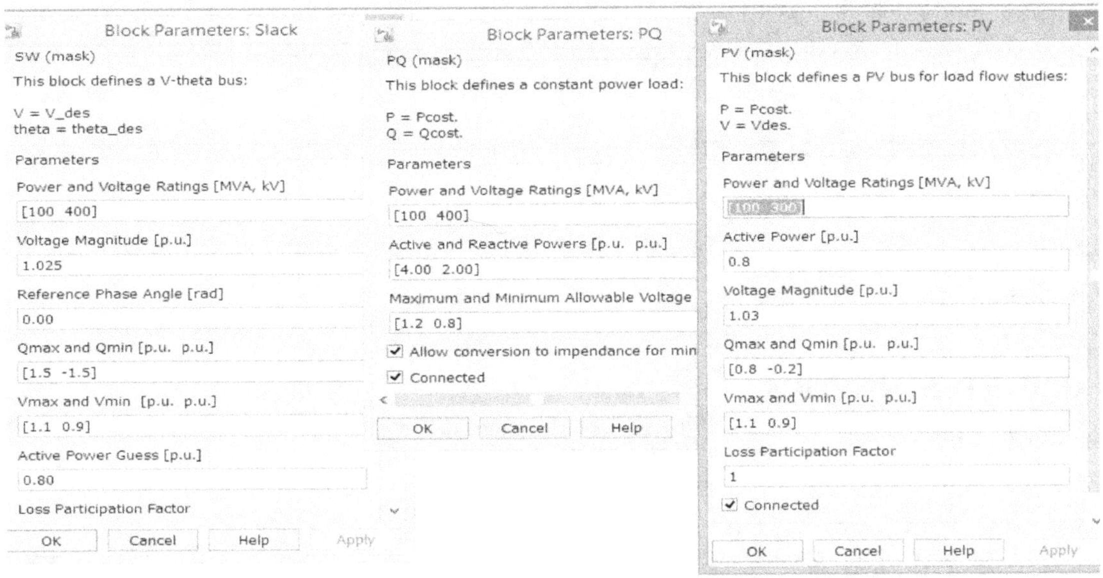

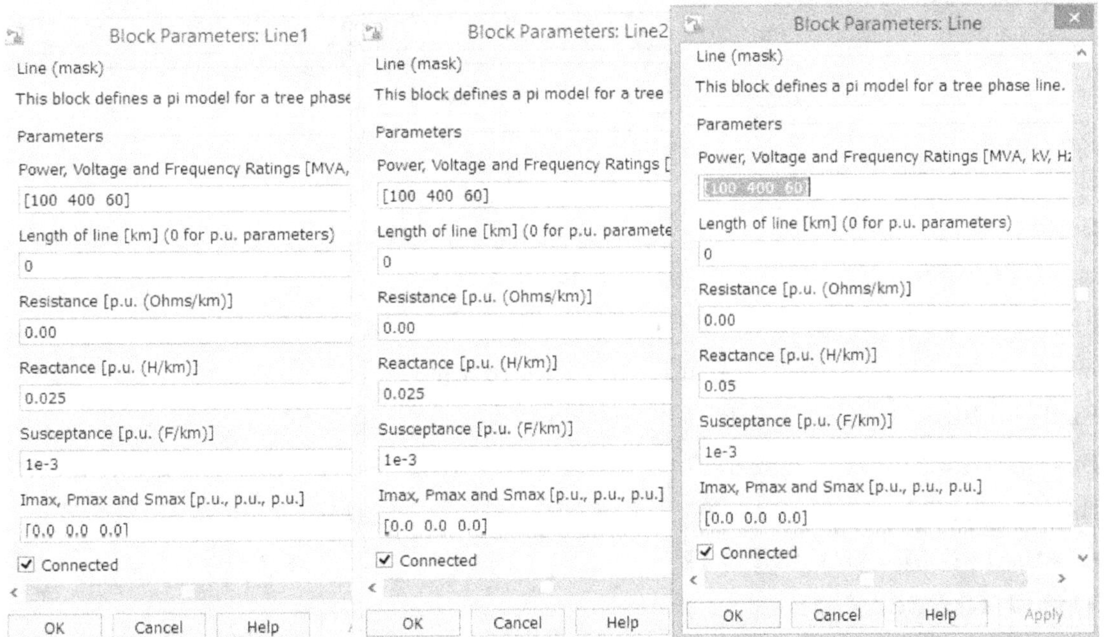

POWER FLOW REPORT

P S A T 2.1.9

Author: Federico Milano, (c) 2002-2013

e-mail: federico.milano@ucd.ie

website: faraday1.ucd.ie/psat.html

File: C:\Program Files\MATLAB\R2014a\bin\psat\tests\q12atry.mdl

Date: 24-Jun-2019 21:01:01

NETWORK STATISTICS

Buses: 3

Lines: 3

Generators: 2

Loads: 1

SOLUTION STATISTICS

Number of Iterations: 3

Maximum P mismatch [p.u.] : 0

Maximum Q mismatch [p.u.]: 0

Power rate [MVA]: 100

POWER FLOW RESULTS

Bus	V [p.u.]	phase [rad]	P gen [p.u.]	Q gen [p.u.]	P load [p.u.]	Q load [p.u.]
Bus1	1.025	0	3.2	1.33	0	0
Bus2	0.99251	-0.06935	0	0	4	2
Bus3	1.03	-0.01801	0.8	1.0048	0	0

LINE FLOWS

From Bus	To Bus	Line	P Flow [p.u.]	Q Flow [p.u.]	P Loss [p.u.]	Q Loss [p.u.]
Bus1	Bus3	1	0.38029	-0.0996	0	0.00629
Bus1	Bus2	2	2.8197	1.4296	0	0.23684
Bus3	Bus2	3	1.1803	0.89894	0	0.09166

GLOBAL SUMMARY REPORT

TOTAL GENERATION

REAL POWER [p.u.]: 4

REACTIVE POWER [p.u.]: 2.3348

TOTAL LOAD

REAL POWER [p.u.]: 4

REACTIVE POWER [p.u.]: 2

TOTAL LOSSES

REAL POWER [p.u.] 0

REACTIVE POWER [p.u.] 0.33479

Example 14: Fig.6.19 shows the single line diagram of a 3 bus power system. The generator is connected at bus 1. The value of voltage at bus 1 is considered as1.155 p.u. the scheduled load at bus 2 and 3 are shown on the diagram. The line impedances are presented in per unit on 100 MVA base. The line charging susceptances are neglected?

Solution:

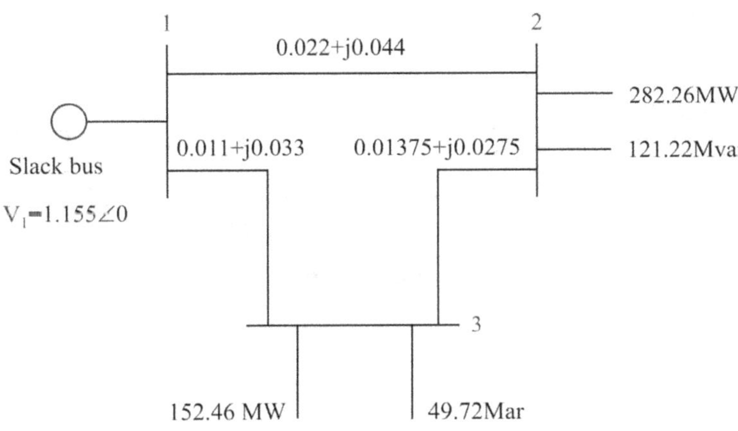

Fig.6.19

Line impedances are converted to admittance:

$$Y_{12} = \frac{1}{0.022+j0.044} = 9.09 - j18.18$$

Similarly, $Y_{13} = 9.09 - j27.27$

$Y_{23} = 14.54 - j29.09$

At the P Q buses, the complex loads expressed in per units are

$$S_2^{sch} = -2.8226 - j1.2122$$

$$S_3^{sch} = -1.5246 - j0.4972$$

Let $V_2^{(0)} = 1.0 + j0.0$ and $V_3^{(0)} = 1.0 + j0.0$

Now, $V_2^{(1)} = \dfrac{\frac{P-JQ}{V} + \Sigma VY}{\Sigma Y}$

$$V_2^{(1)} = \frac{-2.8226 + j1.2122 + (25.03 - j50.08)}{23.63 - j47.27}$$

$$V_2^{(1)} = 1.0286 - j0.0442$$

Now for $V_3^{(1)} = \dfrac{-1.5246+j0.4972+[1.155(9.09-j27.27)+(14.54-j29.09)]}{23.63-j56.36}$

$V_3^{(1)} = 1.055 - j0.02534$

For the second iteration we have

$$V_2^{(2)} = \frac{\frac{-2.8226+j1.2122}{1.0286-j0.0442} + 1.155(9.09 - j18.18) + (1.055 - j0.02534)(14.54 - j29.09)}{23.63 - j47.27}$$

$$V_2^{(2)} = \frac{22.3663 - j50.7593}{23.63 - j47.27}$$

$V_2^{(2)} = 1.0461\text{-}j0.04853$

Again for $V_3^{(2)}$,

$$V_3^{(2)} = \frac{\frac{-1.5246+j0.4972}{1.035+j0.02534} + 1.155(9.09 - j27.27) + (14.54 - j29.09)(1.0286 - j0.0442)}{23.63 - j56.36}$$

$$V_3^{(2)} = \frac{10.4971 - j60.8429}{23.63 - j56.36}$$

$V_3^{(2)} = 1.07297\text{-}j0.048$

For the third iteration we have,

$$V_2^{(3)} = \frac{\frac{-2.8226+j1.2122}{1.0461-j0.04853} + 1.155(9.09 - j18.18) + (= 1.07297 - j0.048)(14.54 - j29.09)}{23.63 - j47.27}$$

$$V_2^{(3)} = \frac{15.5971 - j40.096}{23.63 - j47.27}$$

$V_2^{(3)} = 1.064\text{-}j0.06714$

For $V_3^{(3)}$:

$$V_3^{(3)} = \frac{\frac{-1.5246+j0.4972}{= 1.07297-j0.048} + (10.4998 - j31.4968) + (1.064 - j0.06714)(14.54 - j29.09)}{23.63 - j56.36}$$

$$V_3^{(3)} = \frac{22.9 - j61.8668}{23.63 - j56.36}$$

$V_3^{(3)} = 1.106\text{-}j0.007711$

The power flow solution are:

$V_2 = 1.064j0.06714$

$V_2 = 1.06851 \angle{-0.063}$

$V_3 = 1.106 - j0.00771$

$V_3 = 110606 \angle{-0.00697}$

Example 15: Diagram 6.20 represents the one line drawing of normal 3 bus power system. The generators are connected at buses 1 and 3 respectively. The p.u. value of voltage at bus 1 is considered as 2.1 p.u. and the voltage at bus 3 is set at 2.08 p.u. Also, the real power generation of bus 3 is of 400MW. A load of 400MW and 500Mvar is obtained from bus 2. Line impedences are represented on 100MVA base per unit, and the line charging susceptance is neglected. Obtain the Load flow result by using Gauss Seidel method?

Solution:

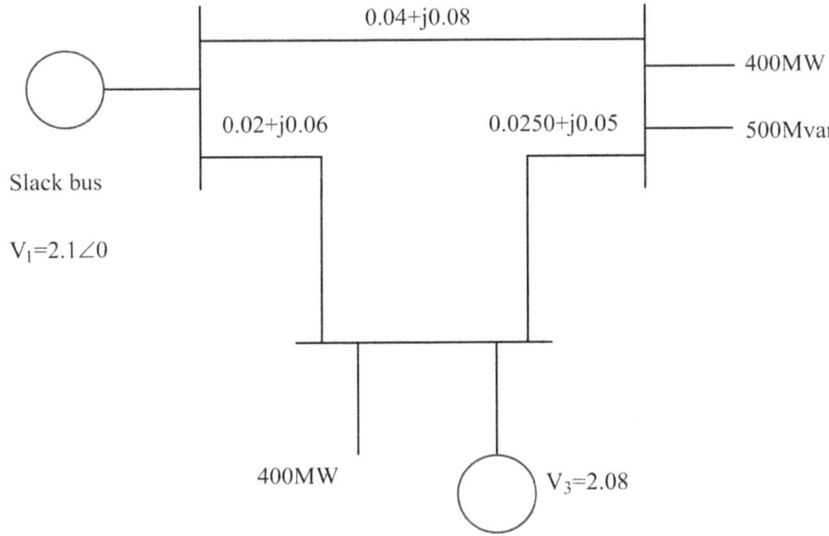

Fig.6.20

One line diagram (Impedance input on 100 MVA base)

Line impedances converted to admittances are

$$Y_{12} = \frac{1}{0.04+j0.08} = 5 - j10$$

Similarly,

$Y_{13} = 5 - j15, \; y_{23} = 8 - j16$

The load and generation expressed in per units are

$$S_2^{sch} = -\frac{400+j500}{100} = -8.0 - j5.0 \text{ p.u.}$$

$P_3^{sch} = 400/100 = 4$ p.u.

Bus 1 is taken as reference bus. Starting from an initial estimate of

$$V_2^{(0)} = 1.0 + j0.0 \qquad\qquad V_3^{(0)} = 2.08 + j0.0$$

$$V_2^{(1)} = \frac{\dfrac{-8+j5}{1+j0.0} + 2.1(5 - 10j) + 2.08(8 - 16j)}{13 - j26}$$

$$V_2^{(1)} = \frac{19.14 - j49.28}{13 - j26} = 1.80984 - j0.16876$$

$$Q_3^{(1)} = -j\{V_3^* [V_3^{(0)} (y_{13} + y_{23}) - (y_{13}v_1 + y_{23}V_2^{(1)})]\}$$

$$Q_3^{(1)} = -j\{2.08[2.08(13 - 31j) - [(5 - 15j)2.1 + (8 - 16j)(1.80984 - 0.16576j)]]\}$$

$$Q_3^{(1)} = -j[(2.08)(4.7615 - 2.673j)]$$

$$Q_3^{(1)} = 5.5598$$

The value of $Q_3^{(1)}$ is used as Q_3^{Sch} for the computation of voltage at bus 3.

The complex voltage at bus 3, denoted by $V_{c3}^{(1)}$ is calculated,

$$V_{c3}^{(1)} = \frac{\dfrac{4 - 5.5598j}{2.08} + (5 - 15j)2.1 + (8 - 16j)(1.80984 - 0.1687j)}{13 - 31j}$$

$$V_{c3}^{(1)} = \frac{24.2016 - 64.48j}{13 - 31j}$$

$$V_{c3}^{(1)} = 2.04736 - 0.07787j$$

Therefore,

$$V_3^{(1)} = 2.04736 - 0.07787j$$

For the second iteration,

$$V_2^{(2)} = \frac{\dfrac{-8+5j}{1.08984 + 0.16876j} + (2.1)(5 - 10j) + (8 - 16j)(2.04736 - 0.07787j)}{13 - 26j}$$

$$V_2^{(2)} = \frac{21.7543 - 51.7344j}{13 - 26j}$$

$$V_2^{(2)} = 1.7376 - j0.18655$$

$$Q_3^2 = -\{V_3^* [V_3^{(1)} (y_{13} + y_{23}) - (y_{13}v_1 + y_{23}V_2^{(2)})]\}$$

$$Q_3^{(2)} = -\{(2.0785 + j0.07787)[(2.0785 - 0.07787)(13 - 31j) - (5 - 15j)2.1$$
$$- (8 - 16j)(1.7376 - j0.18655)$$

$$Q_3^{(2)} = -(15.079 - j14.2663)$$

$Q_3^{(2)} = 14.2663$

The value of $Q_3^{(2)}$ is used as Q_3^{sch} for the computation of voltage at bus 3.

The complex voltage at bus 3, denoted by $V_{c3}^{(2)}$ is calculated,

$$V_{c3}^{(2)} = \frac{\dfrac{4-4.5486j}{2.04736 - 0.07787j} + (2.1)(5 - 15j) + (8 - 16j)(1.7376 - j0.18655)}{13 - 31j}$$

$$V_{c3}^{(2)} = \frac{25.7269 - 65.594j}{13 - 31j}$$

$V_{c3}^{(2)} = 1.0381\text{-}j0.148$

Since, v_3 is held constant at 2.08 p.u, only the imaginary part of $V_{c3}^{(2)}$ is retained,

Therefore $v_3^{(2)} = 1.0381\text{-}j0.148$

The final solution,

$V_2 = 1.7376\text{-}j0.18655$

$V_2 = 1.7724\angle{-}0.107$ p.u.

$V_3 = 1.0381\text{-}j0.148$

$V_3 = 1.060\angle{-}0.1416$ p.u.

$S_3 = 4.0 + j14.263$

p.u.

Example 16: Find the load voltage (bus D)

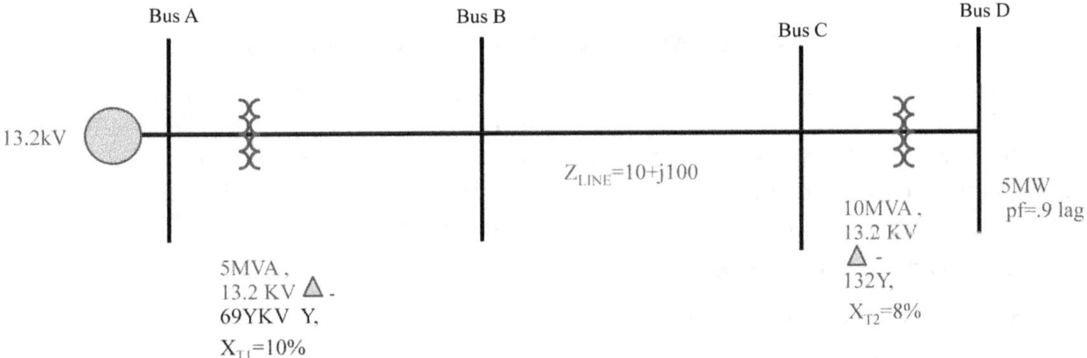

Fig.6.21

$$Y_{BUS} = \begin{bmatrix} 2.88\angle 89 & -2.88\angle 89 \\ -2.88\angle 89 & 2.88\angle 89 \end{bmatrix}$$

Gauss-Seidel Method:

$S_2 = -S_{\text{load,pu}}$, which is the input complex power of Bus D

$$V_2^{i+1} = \frac{1}{Y_{22}}\left[\frac{S_2^*}{V_2^{i*}} - Y_{21}V_1\right]$$

$$V_2^{i+1} = \frac{1}{2.88\angle -89}\left[\frac{-.0556\angle -25.8}{V_2^{i*}} - 2.88\angle 89.1.1\angle 0\right]$$

$$= \frac{-0.193\angle 63.2}{V_2^i} + 1.1$$

1st Iteration:

$$V_2^1 = \frac{-0.193\angle 63.2}{1\angle 0} + 1.1$$

$$= 1.0275\angle -9.6544°$$

$$V_2^2 = \frac{-0.193\angle 63.2}{1.0275\angle -9.6544°}$$

$$= 0.9998\angle -8.6934°$$

$$V_2^3 = \frac{-0.193\angle 63.2}{0.9998\angle -8.6934} + 1.1$$

$$= 1.0003\angle -9.0418°$$

$$V_2^4 = \frac{-0.193\angle 63.2}{1.0003\angle -9.0418°} + 1.1$$

$$= 0.9993\angle -9.0070°$$

$$V_2 = 0.9993\angle -9.0070$$

Power: $S_1 = V_1(Y_{11}^* V_1^* + Y_{12}^* V_2^*)$

$= 1.1 \times (3.168\angle 89 - 2.878\angle 98)$

6.6   Newton-Raphson Method of Load Flow Analysis

By means of Newton-Raphson technique we can work out the Load flow problem. In reality, among the many solution techniques existing in Load flow analysis, the Newton-Raphson technique is regarded as to be most complicated and vital. For a big network Load flow studies the Newton–Raphson method has been established most successful because of several benefits are attributed to the Newton-Raphson (N-R) technique.

Advantages of the N-R Method

1. Acquires less computational time and faster convergence.
2. The Newton-Raphson method is more precise.
3. The Newton-Raphson method is not sensitive to issues such as slack bus choice, regulating transformers etc.
4. The number of iterations involved in this technique is more or less self-governing of the system size i.e. does not depend on the size of the system.

Disadvantages of the N-R Method

1. This method is difficult solution technique.

2. More computations engaged in each iteration results in great computational time per iteration

3. Large requirement of computer memory.

6.6.1 Analytical Concept of N-R Load Flow Technique

Suppose, the unknown variables are $(x_1, x_2, x_3, \ldots\ldots, x_n)$ and the specified variables are $(y_1, y_2, y_3, \ldots\ldots, y_n)$. The relation of the above may be expressed by a set of non-linear equations:

$y_1 = f_1(x_1, x_2, x_3, \ldots\ldots, x_n)$

$y_2 = f_2(x_1, x_2, x_3, \ldots\ldots, x_n)$

$y_3 = f_3(x_1, x_2, x_3, \ldots\ldots, x_n)$

$\ldots \quad \ldots \quad \ldots \quad \ldots \quad \ldots \quad \ldots \quad \ldots$

$y_n = f_n(x_1, x_2, x_3, \ldots\ldots, x_n)$ $\hspace{3cm}$ .....(3.39)

Let us, initial guess solutions of the given non-linear equations are $(x_1^0, x_2^0, x_3^0, \ldots\ldots, x_n^0)$, the zero superscript indicating the zeroth iteration in the course of solving the equations (3.39).

It may be noted that the initial guess solution for the non-linear equations must not be very far away from the original solution, or else, there are probability of the solution diverging to a certain extent than converging and it may not be promising to have a solution irrespective of the computational time.

Normally for N-R method problem the initial guess for a Load flow is considered as a flat voltage profile, i.e., $V_i = (1.0 + j0)$ for i = 1, 2, 3, ..., n apart from the slack bus.

Suppose, $\Delta x_1^0, \Delta x_2^0, \Delta x_3^0, \ldots\ldots \Delta x_n^0$ be the corrections, included to the initial presumed values, give the original solution. Thus,

$$y_1 = f_1(x_1^0 + \Delta x_1^0, x_2^0 + \Delta x_2^0, x_3^0 + \Delta x_3^0, +\ldots\ldots, x_n^0 + \Delta x_n^0) \hspace{2cm} .....(3.40)$$

Newton-Raphson method is solved using Taylor's series and partial derivatives.

Expressing, these equations (Newton-Raphson method is solved using Taylor's series and partial derivatives.) in Taylor's series with respect to the initial guess, we have

$$y_1 = f_1(x_1^0, x_2^0, x_3^0, \ldots\ldots, x_n^0) + \left[\Delta x_1^0 \left(\frac{\partial f_1}{\partial x_1}\right)^0 + \Delta x_2^0 \left(\frac{\partial f_1}{\partial x_2}\right)^0 + \right.$$

$$\Delta x_0^3 \left(\frac{\partial f_1}{\partial x_n}\right)^0 + \ldots.. + \Delta x_n^0 \left(\frac{\partial f_1}{\partial x_n}\right)^0 \left. \right] + \ldots\ldots \text{ higher order terms} \hspace{1.5cm} .....(3.41)$$

where $\left(\frac{\partial f_1}{\partial x_1}\right)^0, \left(\frac{\partial f_1}{\partial x_2}\right)^0, \left(\frac{\partial f_1}{\partial x_n}\right)^0, \ldots, \left(\frac{\partial f_1}{\partial x_n}\right)^0$ are the derivatives of f_1 w.r.t $x_1, x_2, x_3, \ldots\ldots, x_n$ assessed at $(x_1^0, x_2^0, x_3^0, \ldots\ldots, x_n^0)$.

Partial derivatives of second and higher order are ignored. In reality it is this guess that needs the initial solution to be nearer to the ultimate solution.

After making Linearization of all the equations, we obtain a matrix in the form of below:

$$\begin{bmatrix} y_1 - f_1(x_1^0, x_2^0, x_3^0, \ldots, x_n^0) \\ y_2 - f_2(x_1^0, x_2^0, x_3^0, \ldots, x_n^0) \\ y_3 - f_3(x_1^0, x_2^0, x_3^0, \ldots, x_n^0) \\ \ldots\ldots\ldots\ldots\ldots\ldots\ldots \\ y_n - f_n(x_1^0, x_2^0, x_3^0, \ldots, x_n^0) \end{bmatrix} = \begin{bmatrix} \dfrac{\partial f_1}{\partial x_1} \dfrac{\partial f_1}{\partial x_2} \dfrac{\partial f_1}{\partial x_3} \cdots\cdots \dfrac{\partial f_1}{\partial x_n} \\ \dfrac{\partial f_2}{\partial x_1} \dfrac{\partial f_2}{\partial x_2} \dfrac{\partial f_3}{\partial x_3} \cdots\cdots \dfrac{\partial f_2}{\partial x_n} \\ \dfrac{\partial f_3}{\partial x_1} \dfrac{\partial f_3}{\partial x_2} \dfrac{\partial f_3}{\partial x_3} \cdots\cdots \dfrac{\partial f_3}{\partial x_n} \\ \ldots\ldots\ldots\ldots\ldots\ldots\ldots \\ \dfrac{\partial f_n}{\partial x_1} \dfrac{\partial f_n}{\partial x_2} \dfrac{\partial f_n}{\partial x_3} \cdots\cdots \dfrac{\partial f_n}{\partial x_n} \end{bmatrix} \begin{bmatrix} \Delta x_1^0 \\ \Delta x_2^0 \\ \Delta x_3^0 \\ \ldots \\ \ldots \\ \ldots \\ \Delta x_n^0 \end{bmatrix}$$

.....(3.42)

The above expression may be presented as B = J. C (vector form)

Here J indicates Jacobian matrix, a square matrix of the partial derivatives. The solution of the equations requires computation of left hand vector B. The vector B represents the difference of the specified quantities and estimated quantities at $(x_1^0, x_2^0, x_3^0, \ldots, x_n^0)$.

Correspondingly J is estimated at this guess. Solution of the matrix equation offers $(\Delta x_1^0, \Delta x_2^0, \Delta x_3^0, \ldots \Delta x_n^0)$ and the improved approximations of the solution is furnished by

$$x_1^1 = x_1^0 + \Delta x_1^0$$
$$x_2^1 = x_2^0 + \Delta x_2^0$$
$$x_3^1 = x_3^0 + \Delta x_3^0$$
$$x_n^1 = x_n^0 + \Delta x_n^o$$

Reiterating the procedure, with these values, further better estimated values will be obtained. The $(\Delta x_1, \Delta x_2, \Delta x_3, \ldots, \Delta x_n)$ turns into lesser with each iteration and at last the iteration is discontinued when $(\Delta x_1, \Delta x_2, \Delta x_3, \ldots, \Delta x_n)$ are smaller than pre-specified values.

6.6.2. Load Flow Solution with Newton–Raphson (NR) Method

Suppose, for a load flow solution problem real power and voltage magnitude are given for the voltage-controlled buses. The Load flow equation is expressed in polar form.

Consider a power system network portion which is shown in Fig.(6.22). The current incoming to bus i. is I_i.

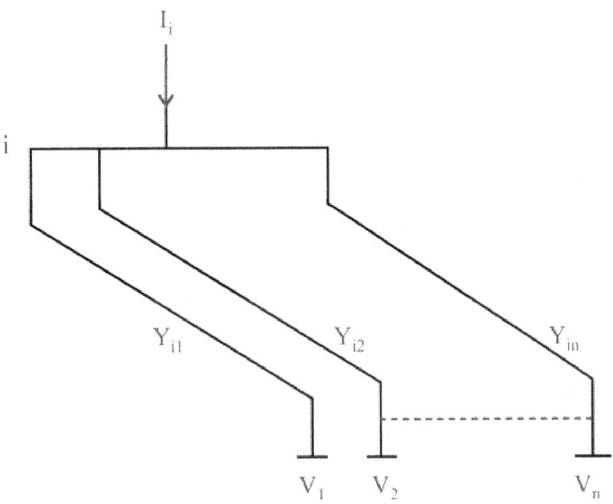

Fig.6.22

The current equation may be obtained with help of following expression as,

$$I_i = \sum_{j=1}^{n} Y_{ij} V_j \qquad \qquad .....(1)$$

The polar representation of the above expression is:

$$I_i = \sum_{j=1}^{n} |Y_{ij}||V_j| \angle \theta_{ij} + \delta_j \qquad \qquad .....(2)$$

The complex power at bus i is may be presented as

$$P_i - jQ_i = V_i^* I_i \qquad \qquad .....(3)$$

From the above equations (2) & (3)

$$P_i - jQ_i = |V| \angle -\delta_i = \sum_{j=1}^{n} |Y_{ij}||V_j| \angle \theta_{ij} + \delta_j. \qquad \qquad ..... (4)$$

Comparing the real and imaginary parts with the left hand side of equation (4),

$$P_i = \sum_{j=1}^{n} |V_i||V_j||Y_{ij}| \cos(\theta_{ij} - \delta_i + \delta_j) \qquad \qquad ..... (5)$$

$$Q_i = -\sum_{j=1}^{n} |V_i||V_j||Y_{ij}| \sin(\theta_{ij} - \delta_i + \delta_j) \qquad \qquad ..... (6)$$

Equation (5) & (6) comprises a set of non linear algebraic equation expressed by the self-determining variables such as bus voltages in per unit, and phase angle in radians. There are two equations for every load bus, represented by (5) & (6) and one equation for every voltage-controlled bus, presented by (5). Expressing (5) & (6) in Taylor's series about the preliminary approximation and ignoring all higher order terms consequences in the subsequent set of linear equations.

$$
\begin{bmatrix} \Delta P_2^{(k)} \\ \vdots \\ \vdots \\ \Delta P_n^{(k)} \\ \Delta Q_2^{(k)} \\ \vdots \\ \vdots \\ \Delta Q_n^{(k)} \end{bmatrix} = \begin{bmatrix} \left(\begin{matrix} \dfrac{\partial P_2^{(k)}}{\partial \delta_2} & \cdots & \dfrac{\partial P_2^{(k)}}{\partial \delta_n} \\ \vdots & \ddots & \vdots \\ \dfrac{\partial P_n^{(k)}}{\partial \delta_2} & \cdots & \dfrac{\partial P_n^{(k)}}{\partial \delta_n} \end{matrix} \right) & \left(\begin{matrix} \dfrac{\partial P_2^{(k)}}{\partial |V_2|} & \cdots & \dfrac{\partial P_2^{(k)}}{\partial |V_n|} \\ \vdots & \ddots & \vdots \\ \dfrac{\partial P_n^{(k)}}{\partial |V_2|} & \cdots & \dfrac{\partial P_n^{(k)}}{\partial |V_n|} \end{matrix} \right) \\ \left(\begin{matrix} \dfrac{\partial Q_2^{(k)}}{\partial \delta_2} & \cdots & \dfrac{\partial Q_2^{(k)}}{\partial \delta_n} \\ \vdots & \ddots & \vdots \\ \dfrac{\partial Q_n^{(k)}}{\partial \delta_2} & \cdots & \dfrac{\partial Q_n^{(k)}}{\partial \delta_n} \end{matrix} \right) & \left(\begin{matrix} \dfrac{\partial Q_2^{(k)}}{\partial |V_2|} & \cdots & \dfrac{\partial Q_2^{(k)}}{\partial |V_n|} \\ \vdots & \ddots & \vdots \\ \dfrac{\partial Q_n^{(k)}}{\partial |V_2|} & \cdots & \dfrac{\partial P_n^{(k)}}{\partial |V_n|} \end{matrix} \right) \end{bmatrix} \begin{bmatrix} \Delta \delta_2^{(k)} \\ \vdots \\ \vdots \\ \Delta \delta_n^{(k)} \\ \Delta \left| V_2^{(k)} \right| \\ \vdots \\ \vdots \\ \Delta \left| V_n^{(k)} \right| \end{bmatrix}
$$

In the above expression, bus 1 is supposed to be the slack bus. The Jacobian matrix provides the linearized connection among little adjustment of Bus voltage angle $\Delta \delta_i^{(k)}$ and Bus voltage magnitude $\Delta \left| V_i^{(k)} \right|$ with the little adjustment in real and reactive power $\Delta P_i^{(k)} \& \Delta Q_i^{(k)}$

Components of the Jacobian matrix are the partial derivatives of (5) & (6), assessed at $\Delta \delta_i^{(k)} \& \Delta \left| V_i^{(k)} \right|$. This may also presented as:

$$
\begin{bmatrix} \Delta P \\ \Delta Q \end{bmatrix} = \begin{bmatrix} J_1 J_2 \\ J_3 J_4 \end{bmatrix} \begin{bmatrix} \Delta \delta \\ \Delta |V| \end{bmatrix} \qquad \qquad \cdots\cdots (7)
$$

In case of voltage regulated bus, the voltages are given. Consequently, if M buses are voltage-regulated, M equations concerning ΔQ & ΔV are the consequent column of the Jacobian matrix is removed. In view of that, there are (N–1) real power constraints and (N–1–M) reactive power constraints, and the Jacobian matrix is of order (2N–2–M) × (2N–2–M). J_1 is of the order (N–1)× (N–1), J_2 is of the order (N–1) × (N–1–M), J_3 is of the order (N–1–M) × (N–1), and J_4 is of the order (N–1–M) × (N–1–M).

The diagonal and off-diagonal components of J_1 are

$$
\frac{\partial P_i}{\partial \delta_i} = \sum_{j \neq i} |V_i| \, |V_j| |Y_{ij}| \sin(\theta_{ij} - \delta_i + \delta_j) \qquad \cdots\cdots (8)
$$

$$
\frac{\partial P_i}{\partial \delta_j} = -|V_i||V_j||Y_{ij}| \sin(\theta_{ij} - \delta_i + \delta_j) j \neq i. \qquad \cdots\cdots (9)
$$

The diagonal and off-diagonal components of J_2 are

$$
\frac{\partial P_i}{\partial |V_i|} = 2|V_i||Y_{ii}| \cos \theta_{ii} + \sum_{j \neq i} |V_j||Y_{ij}| \cos(\theta_{ij} - \delta_i + \delta_j) \qquad \cdots\cdots (10)
$$

$$
\frac{\partial P_i}{\partial |V_j|} = |V_i||Y_{ij}| \cos(\theta_{ij} - \delta_i + \delta_j) j \neq i \qquad \cdots\cdots (11)
$$

The diagonal and off-diagonal components of J_3 are

$$
\frac{\partial Q_i}{\partial \delta_i} = \sum_{j \neq i} |V_i| \, |V_j||Y_{ij}| \cos(\theta_{ij} - \delta_i + \delta_j) \qquad \cdots\cdots (12)
$$

$$\frac{\partial Q_i}{\partial \delta_j} = -|V_i||V_j||Y_{ij}|\cos(\theta_{ij} - \delta_i + \delta_j)j \neq i \qquad\qquad \dots\dots (13)$$

The diagonal and off-diagonal components of J$_4$ are

$$\frac{\partial Q_i}{\partial |V_i|} = -2|V_i||Y_{ii}|\sin\theta_{ii} - \sum_{j\neq i}|V_j||Y_{ij}|\sin(\theta_{ij} - \delta_i + \delta_j) \qquad\qquad \dots\dots (14)$$

$$\frac{\partial Q_i}{\partial |V_j|} = -|V_i||Y_{ij}|\sin(\theta_{ij} - \delta_i + \delta_j)j \neq i \qquad\qquad \dots\dots (15)$$

The terms $\Delta P_i^{(k)}$ and $\Delta Q_i^{(k)}$ are the variation between the scheduled and calculated values which is referred as power residuals, expressed by

$$\Delta P_i^{(k)} = P_i^{sch} - P_i^{(k)} \qquad\qquad \dots\dots (16)$$

$$\Delta Q_i^{(k)} = Q_i^{sch} - Q_i^{(k)} \qquad\qquad \dots\dots (17)$$

The k+1 th approximations for bus voltage are

$$\delta_i^{(k+1)} = \delta_i^{(k)} + \Delta\delta_i^{(k)} \qquad\qquad \dots\dots (18)$$

$$\left|V_i^{(k+1)}\right| = \left|V_i^{(k)}\right| + \Delta\left|V_i^{(k)}\right| \qquad\qquad \dots\dots(19)$$

Numericals on Newton–Raphson Method

$f(x) = x^3 + 9x^2 + 26x + 24 = 0$

Let $x^0 = -5$

$f(x^0) = -125 + 225 - 130 + 24 = -6$

$\dfrac{df(x)}{dx} = 3x^2 + 18x + 26 = 3(-5)^2 + 18(-5) + 26 = 11$

$c - f(x^0) = 0 + 6 = 6$

$x^1 = x^0 + \dfrac{\Delta c^0}{\left[\frac{d(f)}{dx}\right]_0} = -5 + \dfrac{6}{41} = -4.853$

$f(x^1) = (-4.853)^3 + 9(-4.853)^2 + 26(-4.853) + 24 = -4.153$

$\Delta c^1 = c - f(x^1) = 0 + 4.513 = 19.5131$

$\left[\dfrac{d(f)}{dx}\right]_1 = 3(-4.4545)^2 + 18(-4.4545) + 26 = 5.3467$

$x^2 = x^1 + \dfrac{\Delta c^1}{\left[\frac{d(f)}{dx}\right]_1} = -4.853 + \dfrac{1.62259}{5.3467} = -4.15102$

$[f(x)]_2 = (-4.15102)^3 + 9(-4.15102)^2 + 26(-4.15102) + 24 = -0.3739$

$\Delta c^2 = c - f(x^2) = 0.3739$

$\left[\dfrac{d(f)}{dx}\right]_2 = 3(-4.15102)^2 + 18(-4.15102) + 26 = 2.9745$

$$X^3 = x^2 + \frac{\Delta c^2}{\left[\frac{d(f)}{dx}\right]_2} = -4.15102 + \frac{0.3739}{2.9745} = -4.0253$$

Example 17: Using Newton Raphson load flow method, find bus voltages at the end of the end of two iterations for the following 2 bus system. Line reactances are shown in figure. Ignore resistance and line charging. Assume initial voltage at all bus $1.0 < 0^0$. Use 1.0 as acceleration factor. ($V_3 = 1.04 \angle 0°$; $P_{2 \text{ scheduled}} = 4$ p.u.)

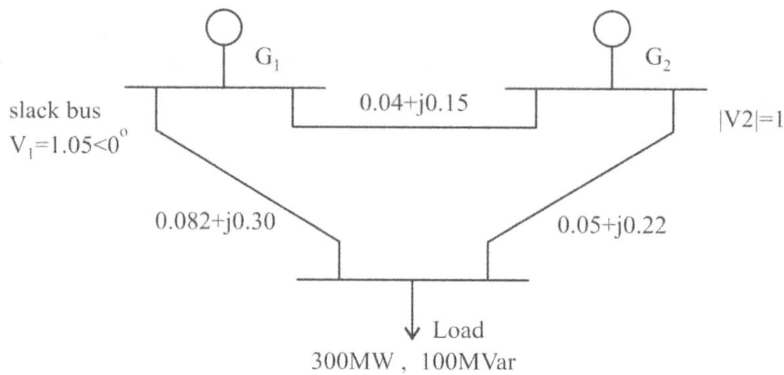

Fig.6.23 Bus Network.

Solution:

$y_{12} = 1/(0.04 + j0.15) = (0.04 - j0.15)/0.0241 = 1.67 - j6.224$

$y_{13} = 1/(0.082 + j0.30) = (0.082 - j0.30)/0.097 = 0.845 - j3.09$

$y_{23} = 1/(0.05 + j0.22) = (0.05 - j0.22)/0.05 = 1 - j4.4$

Now,

$y_{11} = y_{12} + y_{13} = 2.515 - j9.314$

$y_{22} = y_{12} + y_{23} = 2.67 - j10.624$

$y_{33} = y_{13} + y_{23} = 1.845 - j7.49$

$$Y = \begin{bmatrix} 2.515 - j9.314 & -1.67 + j6.224 & -0.845 + j3.09 \\ -1.67 + j6.224 & 2.67 - j10.624 & -1 + j4.4 \\ -0.845 + j3.09 & -1 + j4.4 & 1.845 - j7.49 \end{bmatrix}$$

$$Y = \begin{bmatrix} 9.648\angle - 74.8890 & 3.903\angle105.020 & 3.2035\angle - 74.29433 \\ 3.903\angle - 74.98 & 10.954\angle - 75.890 & 4.5122\angle - 77.19 \\ 3.2035\angle - 74.70 & 4.5122\angle - 77.19 & 7.713\angle - 76.16 \end{bmatrix}$$

Now,

$$\frac{\partial Q_3}{\partial \delta_2} = |V_3| \, |V_2| \, |Y_{32}| \, \text{Cos} \, (\theta_{32} - \delta_3 + \delta_2)$$

$$= 1.04 \times 4.5122 \times \text{Cos} \, (-77.19^0)$$

$$= 1.040$$

$$\frac{\partial Q_3}{\partial \delta_3} = -|V_3|\,|V_1|\,|Y_{31}|\,\text{Cos}\,(\theta_{31} - \delta_3 + \delta_1) - |V_3|\,|V_2|\,|Y_{32}|\,\text{Cos}\,(\theta_{32} - \delta_3 + \delta_2)$$

$$= -1.04 \times 1.05 \times 3.2035 \times \text{Cos}\,(-74.70^0) - 1.04 \times 4.5122 \times \text{Cos}\,(-77.19°)$$

$$= 1.328$$

$$\frac{\partial Q_3}{\partial |V_3|} = |V_1|\,|Y_{31}|\,\text{Sin}\,(\theta_{31} - \delta_3 + \delta_1) + |V_2|\,|Y_{32}|\,\text{Sin}\,(\theta_{32} - \delta_3 + \delta_2) + 2|V_3|\,|Y_{33}|\,\text{Sin}(\theta_{33})$$

$$= 1.05 \times 3.2035 \times \text{Sin}\,(74.70^0) + 1.04 \times 4.5122 \times \text{Sin}\,(-77.19^0)$$

$$+ 2 \times 1.04 \times \text{Sin}\,(-79.04^0)$$

$$= -23.396$$

$$\frac{\partial P_3}{\partial \delta_3} = |V_3||V_1||Y_{31}|\,\text{Sin}(\theta_{31}) + |V_3||V_2||Y_{32}|\,\text{Sin}(\theta_{32})$$

$$= 1.04 \times 1.05 \times 3.2035 \times \text{Sin}\,(-74.70^0) + 1.04 \times 4.5122 \times \text{Sin}\,(-77.19^0)$$

$$= -7.945$$

$$\frac{\partial P_3}{\partial \delta_2} = -|V_3||V_2||Y_{32}|\,\text{Sin}\,(\theta_{32} - \delta_3 + \delta_2)$$

$$= -1.04 \times 4.5122 \times \text{Sin}(-77.19^0)$$

$$= 4.576$$

$$\frac{\partial P_3}{\partial |V_3|} = |V_1|\,|Y_{31}|\,\text{Cos}\,(\theta_{31} - \delta_3 + \delta_1) + |V_2|\,|Y_{32}|\,\text{Cos}(\theta_{32} - \delta_3 + \delta_2) + 2|V_3|\,|Y_{33}|\,\text{Cos}(\theta_{33})$$

$$= 1.05 \times 3.2035 \times \text{Cos}\,(-74.70^0) + 4.5122 \times \text{Cos}\,(-77.19^0) + 2 \times 1.04 \times 7.713 \times \text{Cos}\,(-76.16^0)$$

$$= 5.7166$$

$$\frac{\partial P_2}{\partial \delta_2} = \left|V_2^{(0)}\right|\left|V_1^{(0)}\right||Y_{21}|\,\text{Sin}\,(\theta_{21}) + \left|V_2^{(0)}\right|\left|V_3^{(0)}\right||Y_{23}|\,\text{Sin}(\theta_{23})$$

$$= 1.05 \times 3.903 \times \text{Sin}\,(-74.98^0) + 1.04 \times 4.5112 \times \text{Sin}\,(-77.19^0)$$

$$= -8.52$$

$$\frac{\partial P_2}{\partial \delta_3} = -|V_2||V_3||Y_{23}|\,\text{Sin}\,(\theta_{23} - \delta_2 + \delta_3)$$

$$= -1.04 \times 4.5122 \times \text{Sin}\,(-77.19^0)$$

$$= 4.575$$

$$\frac{\partial P_2}{\partial |V_3|} = 1.05 \times 4.5122 \times \text{Cos}\,(-77.19^0)$$

$$= -1.0504$$

$$P_2 = [|V_2|\,|V_1|\,|Y_{21}|\,\text{Sin}\,(\theta_{21} - \delta_2 + \delta_1) + |V_2|^2\,|Y_{22}|\,\text{Cos}(\theta_{22}) + |V_2|\,|V_3|\,|Y_{23}|\,\text{Cos}\,(\theta_{23} - \delta_2 + \delta_3)]$$

$$P_3 = [|V_3|\,|V_1|\,|Y_{31}|\,\text{Cos}\,(\theta_{31} - \delta_3 + \delta_1) + |V_2||V_3||Y_{32}|\,\text{Cos}\,(\theta_{32} - \delta_3 + \delta_2) + |V_3|^2\,|Y_{33}|\,\text{Cos}\,(\theta_{33})]$$

$$P_{2\text{Calculated}} = [|V_2|\,|V_1|\,|Y_{21}|\,\text{Sin}\,(\theta_{21} - \delta_2 + \delta_1) + |V_2|^2\,|Y_{22}|\,\text{Cos}(\theta_{22}) + |V_2|\,|V_3|\,|Y_{23}|\,\text{Cos}\,(\theta_{23} - \delta_2 + \delta_3)]$$

$P_3 = [|V_3| \, |V_1| \, |Y_{31}| \, \mathrm{Cos} \, (\theta_{31} - \delta_3 + \delta_1) + |V_2| \, |V_3| |Y_{32}| \, \mathrm{Cos} \, (\theta_{32} - \delta_3 + \delta_2) + |V_3|^2 \, |Y_{33}| \, \mathrm{Cos} \, (\theta_{33})]$

$P_{2\text{Calculated}} = [|V_2| \, |V_1| \, |Y_{21}| \, \mathrm{Cos} \, (\theta_{21} - \delta_2 + \delta_1) + |V_2|^2 \, |Y_{22}| \, \mathrm{Cos}(\theta_{22})$

$\qquad + |V_2| \, |V_3| \, |Y_{23}| \, \mathrm{Cos} \, (\theta_{23} - \delta_2 + \delta_3{}^)]$

$= 1.05 \times 3.903 \times \mathrm{Cos} \, (-74.98^0) + 10.954 \times \mathrm{Cos} \, (75.89^0) + 1.04 \times 4.5122 \times \mathrm{Cos} \, (-77.19^0)$

$= 4.772$

$\Delta P_2 = P_{2 \, \text{SCHEDULED}} - P_{2 \, \text{CALCULATED}}$

$= 4 - (4.772)$

$= -0.772$

$P_{3 \, \text{Calculated}} = [|V_3| \, |V_1| \, |Y_{31}| \, \mathrm{Cos} \, (\theta_{21}) + |V_2||V_3||V_{32}| \, \mathrm{Cos} \, (\theta_{32})$

$+ |V_3|^2 \, |Y_{33}| \quad \mathrm{Cos} \, (\theta_{33})$

$= 1.05 \times 1.04 \times 3.2035 \times \cos \, (-74.70) + 1.04 \times 4.5122 - \mathrm{Cos} \, (-77.19)$

$+ 7.63 \times \mathrm{Cos} \, (-76.16) \times (1.05)^2$

$= 2.034$

$\Delta P_3 = P_{3 \, \text{SCHEDULED}} - P_{3 \, \text{CALCULATED}} = 3 - 2.034 = -0.966$

$Q_{3 \, \text{Calculated}} = [|V_3| \, |V_1| \, |Y_{31}| \, \mathrm{Sin} \, (\theta_{31}) + |V_2||V_3||Y_{32}| \, \mathrm{Sin} \, (\theta_{32}) + |V_3|^2 \, |Y_{33}| \, \mathrm{Sin} \, (\theta_{33})]$

$= 1.04 * 1.05 * 3.2035 * \sin 105.2443 + 1.04 * 4.5122 * \sin 102.8 + 1.04^2 \sin -79.04]$

$= 0.895$

$\Delta Q_3 = Q_{3 \, \text{SCHEDULED}} - Q_{3 \, \text{CALCULATED}} = -3 + 0.895 = -2.105$

$$\begin{bmatrix} \Delta P_2 \\ \Delta P_3 \\ \Delta Q_3 \end{bmatrix} = \begin{bmatrix} \dfrac{\partial P_2}{\partial \delta_2} & \dfrac{\partial P_2}{\partial \delta_3} & \dfrac{\partial P_2}{\partial |V_3|} \\[6pt] \dfrac{\partial P_3}{\partial \delta_2} & \dfrac{\partial P_3}{\partial \delta_3} & \dfrac{\partial P_3}{\partial |v_3|} \\[6pt] \dfrac{\partial Q_3}{\partial \delta_2} & \dfrac{\partial Q_3}{\partial \delta_3} & \dfrac{\partial Q_3}{\partial |v_3|} \end{bmatrix} \begin{bmatrix} \Delta \delta_2 \\ \Delta \delta_3 \\ \Delta |v_3| \end{bmatrix}$$

$$\begin{bmatrix} -0.772 \\ 0.966 \\ 3.8893 \end{bmatrix} = \begin{bmatrix} -8.52 & 4.575 & 1.054 \\ 4.576 & -7.950 & 5.7166 \\ 1.040 & 1.328 & -23.396 \end{bmatrix} \begin{bmatrix} \Delta \delta_2 \\ \Delta \delta_3 \\ \Delta |v_3| \end{bmatrix}$$

$$\begin{bmatrix} 0.1170 & 0.0676 & 0.0078 \\ 0.0704 & 0.1656 & -0.0197 \\ -0.0202 & -0.0065 & 0.1709 \end{bmatrix} \begin{bmatrix} 4.1231 \\ -6.72 \\ 3.8893 \end{bmatrix} = \begin{bmatrix} \Delta \delta_2 \\ \Delta \delta_3 \\ \Delta |v_3| \end{bmatrix}$$

$\Delta \delta_2 = 0.0584$

$\Delta \delta_3 = -0.8991$

$\Delta V_3 = -0.6250$

$\delta_2{}^1 = \delta_2{}^0 + \Delta \delta_2{}^0 = 0 + 0.0584 = 0.0584$

$\delta_3{}^1 = 0 - 0.8991 = -0.8991$

$V_3^1 = 1 + 0.6250 = 1.6250$

Solution of the same problem by PSAT [1,2]

POWER FLOW REPORT

P S A T 2.1.9

Author: Federico Milano, (c) 2002-2013

e-mail: federico.milano@ucd.ie

website: faraday1.ucd.ie/psat.html

File: C:\Program Files\MATLAB\R2014a\bin\psat\tests\q17try.mdl

Date: 24-Jun-2019 21:30:16

NETWORK STATISTICS

Buses: 3

Lines: 3

Generators: 2

Loads: 1

SOLUTION STATISTICS

Number of Iterations: 4

Maximum P mismatch [p.u.]: 0

Maximum Q mismatch [p.u.]: 0

Power rate [MVA]: 100

POWER FLOW RESULTS

Bus	V	phase	P gen	Q gen	P load	Q load
	[p.u.]	[rad]	[p.u.]	[p.u.]	[p.u.]	[p.u.]
Bus1	1.05	0	2.2	2.4889	0	0
Bus2	1	-0.07135	0.8	1.7433	0	0
Bus3	0.54928	-0.44413	0	0	2.3571	1.4143

LINE FLOWS

From Bus	To Bus	Line	P Flow	Q Flow	P Loss	Q Loss
[p.u.]	[p.u.]	[p.u.]	[p.u.]			
Bus1	Bus2	1	0.55749	0.2186	0.01302	0.04777

| Bus1 | Bus3 | 2 | 1.6425 | 2.2703 | 0.35622 | 1.5667 |
| Bus3 | Bus2 | 3 | -1.0708 | -0.71063 | 0.27367 | 1.2035 |

LINE FLOWS

From Bus	To Bus	Line	P Flow	Q Flow	P Loss	Q Loss
	[p.u.]	[p.u.]	[p.u.] [p.u.]			
Bus2	Bus1	1	−0.54447	−0.17083	0.01302	0.04777
Bus3	Bus1	2	−1.2863	−0.70363	0.35622	1.5667
Bus2	Bus3	3	1.3445	1.9141	0.27367	1.2035

GLOBAL SUMMARY REPORT

TOTAL GENERATION

REAL POWER [p.u.]: 3

REACTIVE POWER [p.u.]: 4.2322

TOTAL LOAD

REAL POWER [p.u.]: 2.3571

REACTIVE POWER [p.u.]: 1.4143

TOTAL LOSSES

REAL POWER [p.u.] :0.64291

REACTIVE POWER [p.u.] :   2.8179

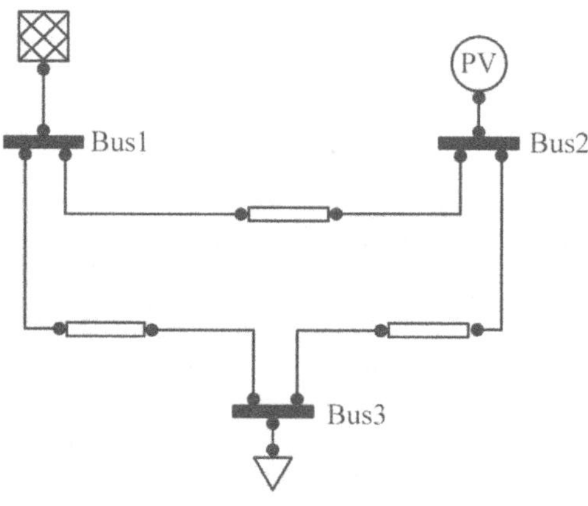

Fig. 6.24

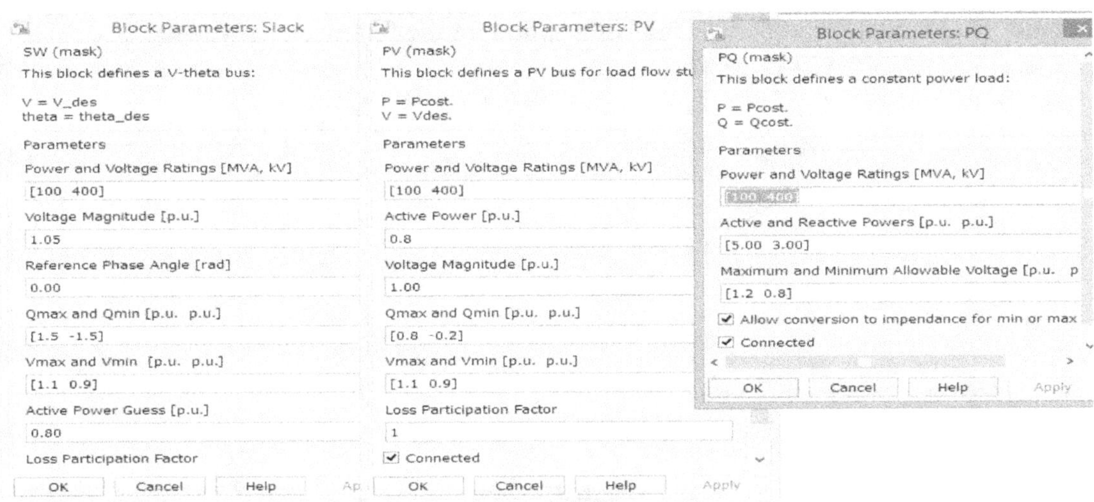

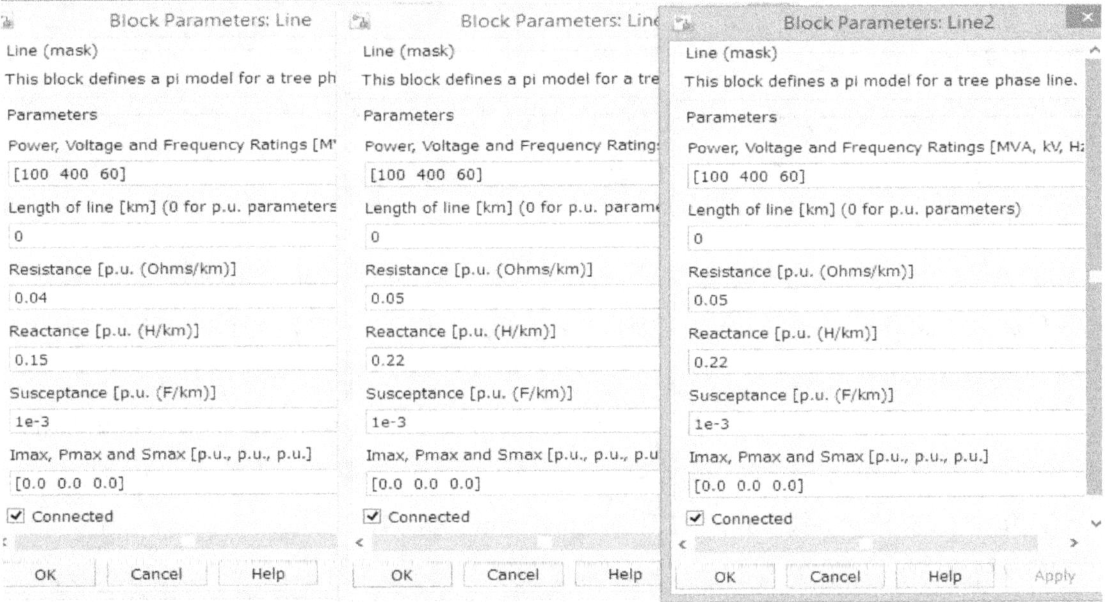

Example 18: Find the Load Flow using NR Method for the following network.

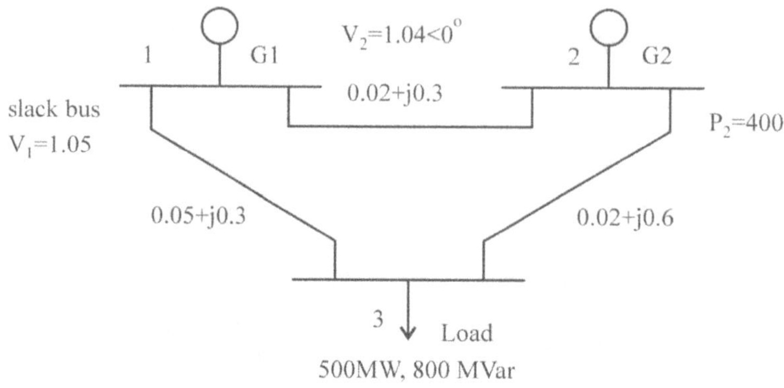

Fig.6.25

$$\frac{\partial Q_3}{\partial \delta_2} = |V_3| |V_2| |Y_{32}| \text{Cos} (\theta_{32} - \delta_3 + \delta_2)$$

$$= (1.04) (1.664) \text{ Cos } (-88.89)$$

$$= 0.0571 \text{ p.u.}$$

$$\frac{\partial Q_3}{\partial \delta_3} = -|V_3| |V_1| |Y_{31}| \text{ Cos } (\theta_{31} - \delta_3 + \delta_1) - |V_3| |V_2| |Y_{32}| \text{ Sin } (\theta_{32} - \delta_3 + \delta_2)$$

$$= -(1.02) (3.288) \text{ Cos } (80.537) - (1.04) (1.664) \text{ Cos}(-88.09)$$

$$= -0.6089 \text{p.u.}$$

$$\frac{\partial Q_3}{\partial |V_3|} = |V_1| |Y_{31}| \text{Sin} (\theta_{31} - \delta_3 + \delta_1) + |V_2||Y_{32}| \text{ Sin } (\theta_{32} - \delta_3 + \delta_2) + 2|V_3| |Y_{33}| \text{ Sin}(\theta_{33})$$

$$= (3.288) (1.02) \text{ Sin } (-80.537) + (1.04) (1.664) \text{ Cos } (-88.09) + 2(4.943) \text{ Sin } (-83.07)$$

$$= -13.064 \text{ p.u.}$$

$$P_3 = [|V_3| |V_1| |Y_{31}| \text{ Cos } (\theta_{31} - \delta_3 + \delta_1) + |V_2| |V_3| |Y_{32}| \text{ Cos } (\theta_{32} - \delta_3 + \delta_2) + |V_3|^2 |Y_{33}| \text{ Cos } (\theta_{33})]$$

$$\frac{\partial P_3}{\partial \delta_3} = |V_3||V_1||Y_{31}| \text{ Sin}(\theta_{31}) + |V_3||V_2||Y_{32}| \text{ Sin}(\theta_{32})$$

$$= (1.02) (3.288) \text{ Sin } (-80.537) + (1.04) (1.664) \text{ Sin } (-88.09)$$

$$= -5.03 \text{ p.u.}$$

$$\frac{\partial P_3}{\partial \delta_2} = -|V_3||V_2||Y_{32}| \text{ Sin } (\theta_{32} - \delta_3 + \delta_2)$$

$$= - (1.04) (1.664) \text{ Sin } (-88.09)$$

$$= -1.729 \text{ p.u.}$$

$$\frac{\partial P_3}{\partial |V_3|} = |V_1| |Y_{31}| \text{ Cos } (\theta_{31} - \delta_3 + \delta_1) + |V_2| |Y_{32}| \text{ Cos } (\theta_{32} - \delta_3 + \delta_2) + 2|V_3| |Y_{33}| \text{ Cos}(\theta_{33})$$

$$= (1.02) (3.288) \text{ Cos } (-80.537) + (1.04) (1.664) \text{ Cos } (-88.09)$$

$$+ 2 (1) (4.943) \text{ Cos } (-83.07) = 1.7976 \text{ p.u.}$$

$P_2 = [|V_2| \, |V_1| \, |Y_{21}| \, \text{Sin} \, (\theta_{21} - \delta_2 + \delta_1) + |V_2|^2 \, |Y_{22}| \, \text{Cos}(\theta_{22}) + |V_2| \, |V_3| \, |Y_{23}| \, \text{Cos} \, (\theta_{23} - \delta_2 + \delta_3)]$

$\dfrac{\partial P_2}{\partial \delta_2} = \left|V_2^{(0)}\right| \left|V_1^{(0)}\right| |Y_{21}| \, \text{Sin}(\theta_{21}) + \left|V_2^{(0)}\right| \left|V_3^{(0)}\right| \left|V_3^{(0)}\right| |Y_{23}| \, \text{Sin}(\theta_{23})$

$\qquad = (1.04) \, (4.949) \, (1.02) \, \text{Sin} \, (\text{-}86.82) + (1.04) \, (1) \, (1.664) \, \text{Sin} \, (\text{-}88.09)$

$\qquad = \text{-}2.788 \text{ p.u.}$

$\dfrac{\partial P_2}{\partial \delta_3} = -|V_2||V_3||Y_{23}| \, \text{Sin} \, (\theta_{23} - \delta_2 + \delta_3)$

$\qquad = - (1.02) \, (1) \, (1.664) \, \text{Sin} \, (\text{-}88.09)$

$\qquad = 1.7296 \text{ p.u.}$

$\dfrac{\partial P_2}{\partial |V_3|} = (1.02) \, (1.664) \, \text{Sin} \, (\text{-}88.09)$

$\qquad = 0.056 \text{ p.u.}$

$P_{2\text{Calculated}} = [|V_2| \, |V_1| \, |Y_{21}| \, \text{Cos} \, (\theta_{21} - \delta_2 + \delta_1) + |V_2|^2 \, |Y_{22}| \, \text{Cos}(\theta_{22}) + |V_2| \, |V_3| \, |Y_{23}| \, \text{Cos} \, (\theta_{23} - \delta_2 + \delta_3^{)}]$

$= [(1.04) \, (1.02) \, (3.325) \, \text{Cos} \, (\text{-}86.189) + (1.04)^2 \, (4.989) \, \text{Cos} \, (\text{-}86.22) + (1.04) \, (1.664) \, \text{Cos} \, (\text{-}88.09)$

$= 0.260 \text{ p.u.}$

$\Delta P_2 = P_{2 \, \text{Scheduled}} - P_{2 \, \text{Calculated}} = 4 - 0.260 = 3.74 \text{ p.u.}$

$P_{3 \, \text{Calculated}} = [|V_3| \, |V_1| \, |Y_{31}| \, \text{Cos}(\theta_{21}) + |V_2||V_3||Y_{32}| \, \text{Cos}(\theta_{32}) + \left|V_3^{(0)}\right|^2 \, |Y_{33}| \, \text{Cos}(\theta_{33})]$

$\qquad = (1.02) \, (3.288) \, \text{Cos} \, (\text{-}80.537) + (1.04) \, (1.664) \, \text{Cos} \, (\text{-}88.09) + (4.943) \, \text{Cos} \, (-83.07)$

$\qquad = 1.204 \text{ p.u.}$

$\Delta P_3 = P_{3 \, \text{Scheduled}} - P_{3 \, \text{Calculated}} = -7 - 1.204 = -8.024 \text{ p.u.}$

$Q_{3 \, \text{Calculated}} = - [|V_3| \, |V_1| \, |Y_{31}| \, \text{Sin} \, (\theta_{31}) + |V_2||V_3||Y_{32}| \, \text{Sin} \, (\theta_{32}) + |V_3|^2 \, |Y_{33}| \, \text{Sin}(\theta_{33})]$

$= - [(3.288) \, (1.02) \, \text{Sin} \, (\text{-}80.537) + (1.04) \, (1.664) \, \text{Sin} \, (\text{-}88.09) + (4.943) \, \text{Sin} \, (-83.07)]$

$= -9.946 \text{ p.u.}$

$\Delta Q_3 = Q_{3 \, \text{Scheduled}} - Q_{3 \, \text{Calculated}} = -3 + 9.943 = 6.943 \text{ p.u.}$

$$\begin{bmatrix} \Delta P_2 \\ \Delta P_3 \\ \Delta Q_3 \end{bmatrix} = \begin{bmatrix} \dfrac{\partial P_2}{\partial \delta_2} & \dfrac{\partial P_2}{\partial \delta_3} & \dfrac{\partial P_2}{\partial |v_3|} \\[2mm] \dfrac{\partial P_3}{\partial \delta_2} & \dfrac{\partial P_3}{\partial \delta_3} & \dfrac{\partial P_3}{\partial |v_3|} \\[2mm] \dfrac{\partial Q_3}{\partial \delta_2} & \dfrac{\partial Q_3}{\partial \delta_3} & \dfrac{\partial Q_3}{\partial |v_3|} \end{bmatrix} \begin{bmatrix} \Delta \delta_2 \\ \Delta \delta_3 \\ \Delta |v_3| \end{bmatrix}$$

$$\begin{bmatrix} \dfrac{\partial P_2}{\partial \delta_2} & \dfrac{\partial P_2}{\partial \delta_3} & \dfrac{\partial P_2}{\partial |v_3|} \\[2mm] \dfrac{\partial P_3}{\partial \delta_2} & \dfrac{\partial P_3}{\partial \delta_3} & \dfrac{\partial P_3}{\partial |v_3|} \\[2mm] \dfrac{\partial Q_3}{\partial \delta_2} & \dfrac{\partial Q_3}{\partial \delta_3} & \dfrac{\partial Q_3}{\partial |v_3|} \end{bmatrix} \begin{bmatrix} \Delta P_2 \\ \Delta P_3 \\ \Delta Q_3 \end{bmatrix} = \begin{bmatrix} \Delta \delta_2 \\ \Delta \delta_3 \\ \Delta |v_3| \end{bmatrix}$$

$$\begin{bmatrix} -2.78 & 1.7296 & 0.056 \\ 1.729 & -5.03 & 1.7976 \\ 0.0576 & -0.608 & -14.854 \end{bmatrix} \begin{bmatrix} 3.74 \\ -8.2028 \\ 6.946 \end{bmatrix} = \begin{bmatrix} \Delta \delta_2 \\ \Delta \delta_3 \\ \Delta |v_3| \end{bmatrix}$$

$\Delta\delta_2 = (-0.2952)\,(3.3523) + (-0.1028)\,(-8.2028) + (-0.013650)\,(6.946) = -0.356$

$\Delta\delta_3 = (0.1023)\,(3.74) + (-0.1650)\,(-8.2028) + (-0.01965)\,(6.946) = 1.5996$

$\Delta V_3 = (0.0028265)\,(3.74) + (-0.005899)\,(-8.2028) + (-0.06803)\,(6.946) = -0.41567$

$\delta_2{}^1 = \delta_2{}^0 + \Delta\delta_2{}^0 = 0 + (-0.356) = -0.356$

$\delta_3{}^1 = 0 + 1.5996 = 1.5996$

$V_3{}^1 = 1 + 0.415676 = 1.415675$

$P_2{}^2{}_{\text{Calculated}} = [|V_2|\,|V_1|\,|Y_{21}|\,\text{Cos}\,(\theta_{21} - \delta_2 + \delta_1) + |V_2|^2\,|Y_{22}|\,\text{Cos}(\theta_{22})$

$\qquad\qquad + |V_2|\,|V_3|\,|Y_{23}|\,\text{Cos}\,(\theta_{23} - \delta_2 + \delta_3)]$

$\qquad\qquad = [(1.04)\,(1.02)\,(4.849)\,\text{Cos}\,(-86.159 + 0.356 + 0) + (1.04)^2\,(2.663)$

$\qquad\qquad \text{Cos}\,(-87.37) + (1.04)\,(1.664)\,\text{Cos}\,(-88.108 + 0.356 - 1.5996)]$

$\qquad\qquad = 0.3259 \text{ p.u.}$

$\qquad\qquad \Delta P_2 = P_{2\,\text{Scheduled}} - P_2{}^2{}_{\text{Calculated}} = 4 - 0.3259 = 0.3259 \text{p.u.}$

$P_{3\,\text{Calculated}} = \quad [|V_3|\,|V_1|\,|Y_{31}|\,\text{Cos}\,(\theta_{21} - \delta_3 + \delta_1) + |V_2||V_3||Y_{32}|$

$\qquad\qquad \text{Cos}\,(\theta_{32} - \delta_3 + \delta_2) + |V_3|^2\,|Y_{33}|\,\text{Cos}\,(\theta_{33})]$

$\qquad\qquad = (1.42)\,(1.02)\,(3.288)\,\text{Cos}\,(-80.546\,\,-1.599+0)$

$\qquad\qquad\quad + (1.42)\,(1.04)\,(1.664)\,\text{Cos}\,(-80.108-1.5996-0.356)$

$\qquad\qquad\quad + (1.42)^2\,(4.944)\,\text{Cos}\,(-83.09)$

$\qquad\qquad = 1.737 \text{ p.u.}$

$\Delta P_3 = P_{3\,\text{Scheduled}} - P_{3\,\text{Calculated}} = -7 - 1.737 = -8.737 \text{ p.u.}$

$Q_{3\,\text{Calculated}} = - [|V_3|\,|V_1|\,|Y_{31}|\,\text{Sin}\,(\theta_{31} - \delta_3 + \delta_1) + |V_2||V_3||Y_{32}|$

$\qquad\qquad \text{Sin}\,(\theta_{32} - \delta_3 + \delta_2) + |V_3|^2\,|Y_{33}|\,\text{Sin}(\theta_{33})]$

$= - [(1.42)\,(3.288)\,(1.02)\,\text{Sin}\,(-80.456\,-\,.5996 + 0) + (1.42)\,(1.04)\,(1.664)\,\text{Sin}\,(-80.108$

$\quad -\,\,1.599 + 0.356) + (1.42)^2\,(1.944)\,\text{Sin}\,(-83.09)]$

$= -\,8.702$ p.u.

$\Delta Q_3 = Q_{3\ \text{Scheduled}} - Q_{3\ \text{Calculated}} = -3 + 8.702 = -5.702$ p.u.

$\dfrac{\partial P_2}{\partial \delta_2} = |V_2||V_1||Y_{21}|\ \text{Sin}\ (\theta_{21} - \delta_2 + \delta_1) + |V_2||V_3||Y_{23}|\ \text{Sin}\ (\theta_{23} - \delta_2 + \delta_3)$

$= (1.04)\,(4.959)\,(1.02)\ \text{Sin}\ (-86.159 + 0.356) + (1.04)\,(1.42)\,(2.663)$

$\quad\text{Sin}\ (-87.37 + 0.365 + 1.5996)$

$= -\,4.978$ p.u.

$\dfrac{\partial P_2}{\partial \delta_3} = -|V_2||V_3||Y_{23}|\ \text{Sin}\ (\theta_{23} - \delta_2 + \delta_3)$

$\qquad = -\,(1.04)\,(1.42)\,(2.663)\ \text{Sin}\ (-87.37 + 0.356 - 1.5996)$

$\qquad = -\,3.920$ p.u.

$\dfrac{\partial P_2}{\partial |V_3|} = (1.04)\,(2.663)\ \text{Cos}\ (-87.37 + 0.356 - 1.5996) = 0.0831$ p.u.

$\dfrac{\partial P_3}{\partial \delta_2} = -|V_2||V_3||Y_{23}|\ \text{Sin}\ (\theta_{23} - \delta_3 + \delta_2)$

$= -\,(1.42)\,(1.04)\,(2.663)\ \text{Sin}\ (-87.37 + 0.356 - 1.5996)$

$= \ 3.920$ p.u.

$\dfrac{\partial P_3}{\partial \delta_3} = |V_3||V_1||Y_{31}|\ \text{Sin}\ (\theta_{31} - \delta_3 + \delta_1) + |V_3||V_2||Y_{32}|\ \text{Sin}\ (\theta_{32} - \delta_3 + \delta_2)$

$= (1.42)\,(1.02)\,(3.488)\ \text{Sin}\ (-80.543 - 1.5996 + 0) + (1.42)\,(1.04)\,(2.663)\ \text{Sin}\ (-87.37 + 0.356$
$+\ 1.5996)$

$\qquad = -\,8.6377$ p.u.

$\dfrac{\partial P_3}{\partial |V_3|} = |V_1||Y_{31}|\text{Cos}\ (\theta_{31} - \delta_3 + \delta_1) + |V_2||Y_{32}|\ \text{Cos}\ (\theta_{32} - \delta_3 + \delta_2) + 2|V_3||Y_{33}|\ \text{Cos}(\theta_{33})$

$= (1.02)\,(3.288)\ \text{Cos}\ (-80.546 - 1.5996 + 0) + (1.04)\,(1.664)\ \text{Cos}\ (-88.108 - 1.5996 + 0.356)$
$+\ 2\,(1.42)\,(4.944)\ \text{Cos}\ (-83.09)$

$= 2.1671$ p.u.

$\dfrac{\partial Q_3}{\partial \delta_2} = -|V_3||V_2||Y_{32}|\ \text{Cos}\ (\theta_{32} - \delta_3 + \delta_2)$

$\qquad = (1.42)\,(1.04)\,(1.664)\ \text{Cos}\ (-88.108 + 1.5996 - 0.356)$

$\qquad = -0.0278$ p.u.

$\dfrac{\partial Q_3}{\partial \delta_3} = |V_3||V_1||Y_{31}|\ \text{Cos}\ (\theta_{31} - \delta_3 + \delta_1) + |V_3||V_2||Y_{32}|\ \text{Sin}\ (\theta_{32} - \delta_3 + \delta_2)$

$\qquad = (1.02)\,(1.42)\,(3.388)\ \text{Cos}\ (-80.546 - 1.5966 + 0) + (1.04)\,(1.42)\,(1.664)$

$\qquad\quad \text{Cos}\ (-88.108 - 1.5996 + 0.35\ )$

$\qquad = 0.67860$ p.u.

$$\frac{\partial Q_3}{\partial |V_3|} = -[|V_1||Y_{31}| \text{ Sin } (\theta_{31} - \delta_3 + \delta_1) + |V_2||Y_{32}| \text{ Sin } (\theta_{32} - \delta_3 + \delta_2) + 2|V_3||Y_{33}|\text{Sin}(\theta_{33})]$$

$$= -[(3.288)(1.02) \text{ Sin } (-80.546 \ 1.5996 + 0) + (1.04)(1.664)$$

$$\text{Sin } (-88.108 - 1.5996 + 0) + 2(1.42)(4.944) \text{ Sin } (-83.09)]$$

$$= 18.9913 \text{ p.u.}$$

$$\begin{bmatrix} \dfrac{\partial P_2}{\partial \delta_2} & \dfrac{\partial P_2}{\partial \delta_3} & \dfrac{\partial P_2}{\partial |v_3|} \\[2mm] \dfrac{\partial P_3}{\partial \delta_2} & \dfrac{\partial P_3}{\partial \delta_3} & \dfrac{\partial P_3}{\partial |v_3|} \\[2mm] \dfrac{\partial Q_3}{\partial \delta_2} & \dfrac{\partial Q_3}{\partial \delta_3} & \dfrac{\partial Q_3}{\partial |v_3|} \end{bmatrix} \begin{bmatrix} \Delta P_2 \\ \Delta P_3 \\ \Delta Q_3 \end{bmatrix} = \begin{bmatrix} \Delta \delta_2 \\ \Delta \delta_3 \\ \Delta |v_3| \end{bmatrix}$$

$$\begin{bmatrix} -4.978 & -3.920 & 0.0831 \\ 3.920 & -8.6377 & 2.1671 \\ -0.0278 & 0.67860 & 18.9913 \end{bmatrix}^{-1} \begin{bmatrix} 3.674 \\ -8.737 \\ 5.702 \end{bmatrix} = \begin{bmatrix} \Delta \delta_2 \\ \Delta \delta_3 \\ \Delta |v_3| \end{bmatrix}$$

$$\Delta \delta_2 = (-0.1483)(3.674) + (0.0667)(-8.737) + (-0.0070)(5.702) = -1.167 \text{ p.u.}$$

$$\Delta \delta_3 = (-0.0668)(3.674) - (-0.0847)(-8.737) + (0.0100)(5.702) = 0.552 \text{ p.u.}$$

$$\Delta V_3 = (0.00223)(3.674) + (0.0031)(-8.737) - +(0.0523)(5.702) = 0.27+ \text{ p.u.}$$

$$\delta_2{}^2 = \delta_2 + \Delta \delta_2 = -0.356 - 1.167 = 1.523$$

$$\delta_3{}^2 = 1.5996 - 0.552 = 2.516$$

$$9666V_3{}^2 = 1.415676 + 0.279 = 1.695$$

Example 19: Load flow study using Newton Raphson for a 3-bus system.

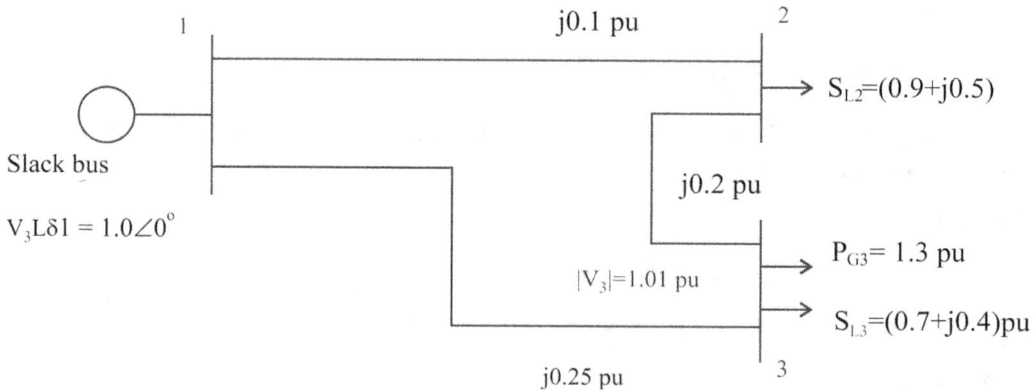

Fig.6.26

Objective: Determine the unknown states of system using Newton Raphson method.

Solution: Known states = $[\delta_1, V_1, V_3]$

Unknown states = $[\delta_2, \delta_3, V_2]$

$$Y_{Bus} = \begin{bmatrix} -j14 & j10 & j4 \\ j10 & -j15 & j5 \\ j4 & j5 & -j9 \end{bmatrix}$$

General power flow equation:

$$P_k = \sum_{i=1}^{n} Y_{(Ki)} V_{(i)} V_{(k)} \cos[\theta_{ik} + \delta_i - \delta_k]$$

$$Q_k = -\sum_{i=1}^{n} Y_{(Ki)} V_{(i)} V_{(k)} \sin[\theta_{ki} + \delta_i - \delta_k]$$

$$F(P_2) = 0 = Y_{21}V_1V_2\cos(\theta_{21}+ \delta_1 - \delta_2) + Y_{22}V_2^2\cos(\theta_{22}) + Y_{23}V_3V_2\cos(\theta_{23} + \delta_3 - \delta_2) - P_2 = 0$$

$$F(P_3) = 0 = Y_{31}V_1V_3\cos(\theta_{31}+ \delta_1 - \delta_3) + Y_{32}V_2V_3\cos(\theta_{32} + \delta_2 - \delta_3) + Y_{33}V_3^2\cos(\theta_{33}) - P_3 = 0$$

$$F(Q_2) = 0 = -Y_{21}V_1V_2\sin(\theta_{21}+ \delta_1 - \delta_3) - Y_{22}V_2^2\sin(\theta_{22}) - Y_{23}V_3V_2\sin(\theta_{23} + \delta_2 - \delta_3) - Q_2 = 0$$

$$[J] = \begin{bmatrix} \dfrac{\partial fP_2}{\partial\delta_2} & \dfrac{\partial fP_2}{\partial\delta_3} & \dfrac{\partial f(P_2)}{\partial|v_3|} \\[2ex] \dfrac{\partial f(P_3)}{\partial\delta_2} & \dfrac{\partial f(P_3)}{\partial\delta_3} & \dfrac{\partial f(P_3)}{\partial|v_3|} \\[2ex] \dfrac{\partial f(Q_3)}{\partial\delta_2} & \dfrac{\partial f(Q_3)}{\partial\delta_3} & \dfrac{\partial f(Q_3)}{\partial|v_3|} \end{bmatrix}$$

$$J_{11} = Y_{21}V_1V_2\sin(\theta_{21}+\delta_1 - \delta_2) + Y_{23}V_3V_2\sin(\theta_{23} + \delta_3 - \delta_2)$$

$$= 1.0 \times 1 \times 10\,\sin(90) + 5 \times 1.01 \times 1\,\sin(90) = 15.05$$

$$J_{12} = -Y_{23}V_3V_2\sin(\theta_{23}+ \delta_3 - \delta_2)$$

$$= -5 \times 1.01 \times 1\sin(90) = -5.05$$

$$J_{13} = Y_{21}V_1\cos(\theta_{21}+ \delta_1 - \delta_2)+2\,Y_{22}V_2\cos(\theta_{22}) + Y_{23}V_3\cos(\theta_{23} + \delta_3 - \delta_2)$$

$$= 1.0 \times 10\,\cos(90)+2 \times 15 \times 1\,\cos(-90) + 5 \times 1.01\,\cos(90) = 0$$

$$J_{21} = -Y_{32}V_2V_3\sin(\theta_{32}+ \delta_2 - \delta_3)$$

$$= -1 \times 1.01 \times 5\sin(90) = -5.05$$

$$J_{22} = Y_{31}V_1V_3\sin(\theta_{31}+ \delta_1 - \delta_3)+Y_{32}V_2V_3\sin(\theta_{32} + \delta_2 - \delta_3)$$

$$= 4 \times 1.01 \times 1.0\,\sin(90)+5 \times 15 \times 1\,\sin(90) = 9.09$$

$$J_{23} = Y_{32}V_3\cos(\theta_{32} + \delta_2 - \delta_3)$$

$$= Y_{32}V_3\cos(90) = 0$$

$$J_{31} = Y_{21}V_1V_2\cos(\theta_{21}+ \delta_1 - \delta_2) + Y_{23}V_3V_2\cos(\theta_{23}+ \delta_3 - \delta_2)$$

$$= Y_{21}V_1V_2\cos(90) + Y_{23}V_3V_2\cos(90) = 0$$

$$J_{32} = -Y_{23}V_3V_2\cos(\theta_{23} + \delta_3 - \delta_2)$$

$$= -Y_{23}V_3V_2\cos(90) = 0$$

$$J_{33} = -Y_{21}V_1\sin(\theta_{21}+ \delta_1 - \delta_2) - 2Y_{22}V_2\sin(\theta_{22}) - Y_{23}V_3\sin(\theta_{23}+ \delta_3 - \delta_2)$$

$$= -10 \times 1.0\sin(90) - 2 \times 15 \times 1\sin(-90) - 5 \times 1.01\sin(90) = 14.95$$

$$\begin{bmatrix} \Delta\delta_2 \\ \Delta\delta_3 \\ \Delta|V_3| \end{bmatrix} = \begin{bmatrix} \dfrac{\partial P_2}{\partial\delta_2} & \dfrac{\partial P_2}{\partial\delta_3} & \dfrac{\partial P_2}{\partial|v_3|} \\[6pt] \dfrac{\partial P_3}{\partial\delta_2} & \dfrac{\partial P_3}{\partial\delta_3} & \dfrac{\partial P_3}{\partial|v_3|} \\[6pt] \dfrac{\partial Q_3}{\partial\delta_2} & \dfrac{\partial Q_3}{\partial\delta_3} & \dfrac{\partial Q_3}{\partial|v_3|} \end{bmatrix}^{-1} \times \begin{bmatrix} \Delta P_2 \\ \Delta P_3 \\ \Delta Q_3 \end{bmatrix}$$

$$\begin{bmatrix} \Delta\delta_2 \\ \Delta\delta_3 \\ \Delta|V_3| \end{bmatrix} = \begin{bmatrix} \dfrac{\partial P_2}{\partial\delta_2} & \dfrac{\partial P_2}{\partial\delta_3} & \dfrac{\partial P_2}{\partial|v_3|} \\[6pt] \dfrac{\partial P_3}{\partial\delta_2} & \dfrac{\partial P_3}{\partial\delta_3} & \dfrac{\partial P_3}{\partial|v_3|} \\[6pt] \dfrac{\partial Q_3}{\partial\delta_2} & \dfrac{\partial Q_3}{\partial\delta_3} & \dfrac{\partial Q_3}{\partial|v_3|} \end{bmatrix}^{-1} \times \begin{bmatrix} \Delta P_2 \\ \Delta P_3 \\ \Delta Q_3 \end{bmatrix}$$

$$[Y] = \begin{bmatrix} 14\angle-90 & 10\angle90 & 4\angle90 \\ 10\angle90 & 15\angle-90 & 5\angle90 \\ 4\angle90 & 5\angle90 & -9\angle-90 \end{bmatrix}$$

$\Delta\delta_2\delta_2 f(P_2)$

$\Delta\delta_3 = \delta_3 - J^{-1} \times f(P_3)$

$\Delta\delta_2\delta_2 f(P_2)$

$V_2 V_2 f(Q_2)$

$J_{11} = 15.05,\ J_{12} = -5.05,\ J_{13} = 0,\ J_{21} = -5.05,\ J_{22} = 9.09,\ J_{23} = 0,\ J_{31} = 0,\ J_{32} = 0,\ J_{33} = 14.95$

$F(P_2) = 0 = 1.0 \times 1 \times 14\cos(90) + 15 \times 1\cos(-90) + 5 \times 1.01 \times 1\cos(90) = 0$

$F(P_3) = 0 = Y_{31}V_1V_3\cos(\theta_{31} + \delta_1 - \delta_3) + Y_{32}V_2V_3\cos(\theta_{32} + \delta_2 - \delta_3) + Y_{33}V_3^2\cos(\theta_{33}) = 0$

$F(Q_2) = 0 = -10 \times 1.0 \times 1\sin(90) - 15 \times 1\sin(-90) - 15 \times 1 \times -5\sin(90) = 0.05$

$\Delta P_2 = P_2\ \text{Scheduled} - P_2\ \text{Calculated}$

$= -0.9 - 0 = -0.9$

$\Delta P_3 = P_3\ \text{Scheduled} - P_3\ \text{Calculated}$

$= 0.6 - 0 = 0.6$

$\Delta Q_3 = Q_3\ \text{Scheduled} - Q_3\ \text{Calculated}$

$= 0.4 + 0.05 = 0.45$

1st Iteration:

$$\begin{bmatrix} 15.05 & -5.05 & 0 \\ -5.05 & 9.09 & 0 \\ 0 & 0 & 14.95 \end{bmatrix}^{-1} \begin{bmatrix} -0.9 \\ 0.6 \\ -0.45 \end{bmatrix} = \begin{bmatrix} \Delta\delta_2 \\ \Delta\delta_3 \\ \Delta|v_2| \end{bmatrix}$$

$$\begin{bmatrix} 0.0817 & 0.0454 & 0 \\ 0.0454 & 0.1352 & 0 \\ 0 & 0 & 0.0669 \end{bmatrix} \begin{bmatrix} -0.9 \\ 0.6 \\ -0.45 \end{bmatrix} = \begin{bmatrix} \Delta\delta_2 \\ \Delta\delta_3 \\ \Delta|v_2| \end{bmatrix}$$

$\Delta\delta_2 = 0.10077$

$\Delta\delta_3 = 0.04026$

$V_2 = -0.0301$

2nd Iteration:

$F(P_2) = 0.02647, \ F(P_3) = 0.001772, \ F(Q_2) = -0.0489897$

$J_{11} = 15.0493, \ J_{12} = -5.0499, \ J_{13} = 0.02292, \ J_{21} = -5.0499, \ J_{22} = 9.0899, \ J_{23} = -0.012430,$

$J_{31} = -0.03359, \ J_{32} = -0.00533, \ J_{33} = 14.850$

$\Delta P_2 = P_2 \text{ Scheduled} - P_2 \text{ Calculated}$

$= -0.9 - 0.02647 = -0.92647$

$\Delta P_3 = P_3 \text{ Scheduled} - P_3 \text{ Calculated}$

$= 0.6 - 0.001772 = 0.5982$

$\Delta Q_3 = Q_3 \text{ Scheduled} - Q_3 \text{ Calculated}$

$= 0.4 + 0049897 = 0.449897$

$$\begin{bmatrix} 15.0493 & -5.0499 & 0.02292 \\ -5.0499 & 9.0899 & -0.012430 \\ 0.0335 & -0.00533 & 14.850 \end{bmatrix}^{-1} \begin{bmatrix} -026.9 \\ 0.598 \\ -0.4499 \end{bmatrix} = \begin{bmatrix} \Delta\delta_2 \\ \Delta\delta_3 \\ \Delta|v_2| \end{bmatrix}$$

$$\begin{bmatrix} 0.0816 & 0.0453 & -0.0001 \\ 0.0453 & 0.1352 & 0.0 \\ -0.0002 & -0.0001 & 0.0673 \end{bmatrix} \begin{bmatrix} -026.9 \\ 0.598 \\ -0.4499 \end{bmatrix} = \begin{bmatrix} \Delta\delta_2 \\ \Delta\delta_3 \\ \Delta|v_2| \end{bmatrix}$$

$L\delta_2 = -0.0485$

$L\delta_3 = \ \ 0.0389$

$V_2 \ \ = \ \ \ 0.03$

Example 20: Find using Gauss Seidel and Newton Raphson Method $V_2^{(2)}$, $\delta_2^{(2}$, $\delta_3^{(2)}$

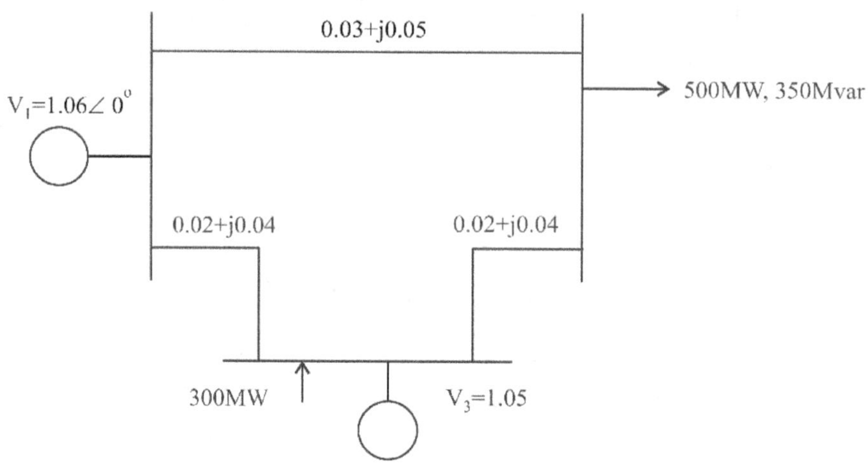

Fig.6.27

Gauss-Siedel Method

Y12=1\(0.03+0.05j) =8.82353-j14.70589

Y13=1\(0.02+j0.04) =10-j20

Y23=1\(0.02+j0.04) =10-j20

$$Y_{Bus} = \begin{bmatrix} 18.82353 - j34 & -8.82353 + j14.70589 & -10 + j20 \\ -8.82353 + j14.70589 & 18.82358 - j34.70589 & -10 + j20 \\ -10 + j20 & -10 + j20 & 20 - j40 \end{bmatrix}$$

2^{sch} =-(500+j350) \100=-5-j3.5

P_3^{sch} =300/100=3

$V_2^{(1)}$ =((-5+j3.5) +(8.82353-j14.70589) (1.06) +(10-j20) (1.05))

\(8.82353-j14.70589+10-j20)

=0.9160377-j0.068868

$Q_3^{(1)}$ =-im (1.05(1.05(10-j20+10-j20)-1.06(10-j20) -(10-j20) (0.916038-j0.068868)

=1.88

$V_3^{(1)}$ =((3-j1.88) \1.05) +(10-j20) (1.06) +(10-j20) (0.916038-j0068863)) \(20-j40)

=1.05239952+j0.004806

$$e_3^{(1)} = \sqrt{(1.05)^2 - (0.004806)^2}$$

$V_3^{(1)}$ =1.05-j0.004806

2nd iteration

$$V_2^{(2)} = \frac{\frac{(-5j3.5)}{(0.916038+j0.068863)}+((8.82853-j14.70589)(10-j20)(1.05-j0.004806)}{(8.82353-j14.70589+10-j20)}$$

$$=0.913826\text{-}j0.072647=0.09167 \text{ L-}0.07933$$

$Q_3^{(2)}$ =[(1.05+j0.004806) ((1.05-j0.004806) (10-j20+10-j20)-1.06(10-j20) - (0.913826-j0.072647) (10-j20))]

=1.69

$$V_3^{(2)} = \frac{\frac{3-j1.69}{1.05+j0.004806}+(10-j20)(1.06)+(10-j20)(0.93826-j0.072647)}{(20-j40)}$$

$$=1.0478\text{+}j0.004451$$

$$e_3^{(2)} = \sqrt{(1.05)^2-(0.004451)^2}=1.0499=1.05$$

$$V_3^{(2)}=1.05\text{-}j0.004451=1.05 \text{ L -}0.0439$$

Hence

$V_2^{(2)}=0.09167$

$\delta_2^{(2)}=\text{-}0.07933$

$\delta_3^{(2)}=-0.0439$

Newton Raphson Method

$Y_{12}=1\backslash(0.03+0.05j)=8.82353\text{-}j14.70589$

$Y_{13}=1\backslash(0.02+j0.04)=10\text{-}j20$

$Y_{23}=1\backslash(0.02+j0.04)=10\text{-}j20$

$$Y_{Bus} = \begin{bmatrix} 18.82353-j34 & -8.82353+j14.70589 & -10+j20 \\ -8.82353+j14.70589 & 18.82358-j34.70589 & -10+j20 \\ -10+j20 & -10+j20 & 20-j40 \end{bmatrix}$$

$P_i=\sum_{j=1}^{n}\left|V_iY_{ij}V_j\right|\cos(\theta_{ij}+\delta_j-\delta_i)$

$Q_i=\sum_{j=1}^{n}\left|V_iY_{ij}V_j\right|\sin(\theta_{ij}+\delta_j-\delta_i)$

Taking initial guess $V_2^{(0)}=1$ and $\delta_2^{(0)}=0$ and $\delta_3^{(0)}=0$

$\frac{\partial P_2}{\partial \delta_2}=(1\times17.14986\times1.06)\sin(59.0306)+(1.05\times1.05\times1\times22.3606)\sin(63.4349)=37.815$

$\frac{\partial P_2}{\partial \delta_3}=(1.05\times22.3606\times1.06)\sin(63.4349)+(1.05\times22.3606\times1)\times\sin(63.4349)=43.25$

$\frac{\partial P_2}{\partial V_2} = (1.06 \times 17.14986 \times \cos(59.0363) + 2 \times (39.48194\cos(-61.5258)) + (1.04 \times 22.3606 \times \cos(63.4349)) = 17.894$

$\frac{\partial P_3}{\partial \delta_2} = -(1.05 \times 22.3606) \times \sin(63.4349) = -20.9$

$\frac{\partial P_3}{\partial \delta_3} = (1.05 \times 1.06 \times 22.3606) \times \sin(63.4349) + (1.05 \times 22.3606 \times \sin(63.4349)) = 42.9$

$\frac{\partial P_3}{\partial V_2} = (1.05 \times 2.3606) \times \cos(63.4349) = 10.499$

$\frac{\partial Q_2}{\partial \delta_2} = (1 \times 1.06 \times 17.14986) \times \cos(59.0306) + (1 \times 1.05 \times 22.3606) \times \cos(63.4349) = 20.0431$

$\frac{\partial Q_2}{\partial \delta_3} = -(1 \times 1.05 \times 22.3606) \times \cos(63.4349) = -10.4$

$\frac{\partial Q_2}{\partial V_2} = -(1.06 \times 17.14986 \times \sin(59.063)) - 2 \times (1 \times 39.48194) \times \sin(-61.5258) - (1.05 \times 22.3606)\sin(63.4349) = 32.82$

$P_2 = (1 \times 17.14986 \times 1.06) \times \cos(59.0363) + (39.48194 \times 1) \times \cos(-61.5258) + (1 \times 22.3606 \times 1.05) \times \cos(63.4349)$

$\qquad = 1.0290$

$P_3 = (1.05 \times 22.3606 \times 1.06) \times \cos(63.4349) + (1.05 \times 22.3606 \times 1)$

$\qquad \times \cos(63.4349) + (1.05 \times 44.72136 \times 1.05) \times \cos(63.4349)$

$\qquad = 0.4202$

$Q_2 = (1 \times 17.14986 \times 1.06) \times \sin(59.0363) + (39.48194 \times 1) \times \sin(-61.5258)$

$\qquad + (22.3606 \times 1.05) \times \sin(63.4349)$

$\qquad = 1.88246$

Now power mismatch,

$\Delta P_2^{(0)} = -5 - (-1.0290) = -3.971$

$\Delta P_3^{(0)} = 3 - (0.42050) = 2.57944$

$\Delta Q_2^{(0)} = -3.5 - (1.88246) = -5.38246$

Now:

$$\begin{bmatrix} -3.971 \\ 2.57944 \\ -5.38246 \end{bmatrix} = \begin{bmatrix} 37.815 & 43.25 & 17.894 \\ -20.9 & 42.9 & 10.49 \\ 20.0431 & -10.4 & 32.82 \end{bmatrix} \begin{bmatrix} \Delta\delta_2^{01} \\ \Delta\delta_3^{01} \\ \Delta V_2^{01} \end{bmatrix}$$

Or,
$$\begin{bmatrix} \Delta\delta_2^{01} \\ \Delta\delta_3^{01} \\ \Delta V_2^{01} \end{bmatrix} = \begin{bmatrix} -0.0466 \\ 0.0316 \\ -0.2019 \end{bmatrix}$$

Now:

$V_2^{(1)} = 1 + (-0.0919) = 0.90821$

$\delta_2^{(1)} = -0.1001$

$\delta_3^{(1)} = 0.0754$

References

1. F. Milano, "An open source power system analysis toolbox," in *IEEE Transactions on Power Systems*, vol. 20, no. 3, pp. 1199-1206, Aug. 2005. doi: 10.1109/TPWRS.2005.851911

2. F. MIlano, "Power System Modelling and Scripting," Springer, London, 2010."

Exercise

Questions

Objective Type

Fill in the Blanks

1. In a load flow problem solved by Newton-Raphson method with polar coordinates, the size of the Jacobian is 100×100. If there are 20 PV buses in addition to PQ buses and a slack bus, the total number of buses in the system is _________.   (GATE)

2. A 1000 × 1000 bus admittance matrix for an electric power system has 8000 non-zero elements. The minimum number of branches (transmission lines and transformers) in this system are ______ (up to 2 decimal places).   (GATE)

3. A 10-bus power system consists of four generator buses indexed as G1, G2, G3, G4 and six load buses indexed as L1, L2, L3, L4, L5, L6. The generator bus G1 is considered as slack bus, and the load buses L3 and L4 are voltage controlled buses. The generator at bus G2 cannot supply the required reactive power demand, and hence it is operating at its maximum reactive power limit. The number of non-linear equations required for solving the load flow problem using Newton-Raphson method in polar form is ____________.

                (GATE)

4. A 183-bus power system has 150PQ buses and 32 PV buses. In the general case, to obtain the load flow solution using Newton-Raphson method in polar coordinates, the minimum number of simultaneous equations to be solved is ___________ .　(GATE)

5. A power system consists of 300 buses out of which 20 buses are generator buses, 25 buses are the ones with reactive power support and 15 buses are the ones with fixed shunt capacitors. All the other buses are load buses. It is proposed to perform a load flow analysis for the system using Newton-Raphson method. The size of the Newton-Raphson Jacobian matrix is ----------------------- 　(GATE)

6. In load-flow analysis, a voltage-controlled bus is treated as a load bus in subsequent iteration for a _________ limit is violated.　(GATE)

Chapter 7

Economic Load Dispatch

7.1 Introduction of Economic Load Dispatch

The rising electricity requirement and diminishing energy resources have demanded the best possible, utilize of accessible fossil fuel reserves. Subsequently, the fossil fuels are continually rising in cost. Proper arrangement of existing generating units to convene the load requirement is a vital work of a power system engineer to fulfill the economic criteria of the customer as well as to obtain the revenues on the capital investment for the generating company. The problem of allocation of the total load requirement between existing working generators cost-effectively and also fulfilling different system limitations all together is identified as Economic Load Dispatch.

7.2 Economic Scheduling

Economic scheduling gives the allocation of the load among the various units in service in such a way that total cost of generation is minimum.

Unit commitment problem decides which units are kept ON and which unit are kept OFF to supply a given load demand for particular period of time.

Incremental Cost

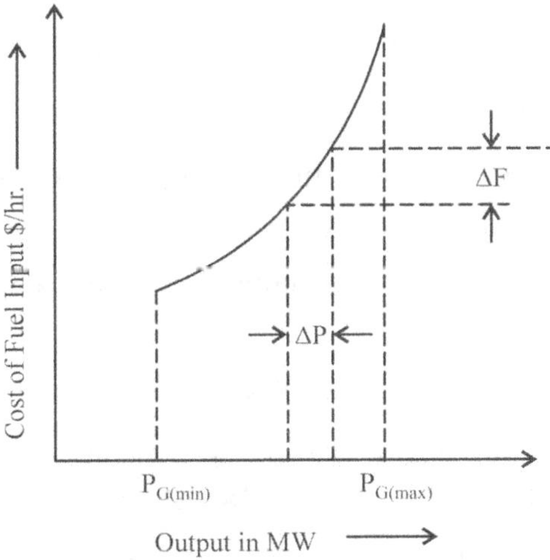

Fig.7.1 Incremental Cost Curve.

Fig.7.1 illustrates the incremental cost curve. From the input-output curve a small change in input (ΔF) in terms of \$/MWhr. and its subsequent change in output (ΔP_G) in terms of MW are taken. The Incremental Fuel Cost (IFC) is termed as the ratio of ΔInput (ΔF) to its corresponding ΔOutput (ΔP_G). Hence,

$$\text{IFC} = (\Delta \text{Input})/(\Delta \text{Output}) = \Delta F/\Delta P_G$$

7.3　Formulation of Economic Load Dispatch Problem

The present study for Economic Load Dispatch is exploited to solve the operation and planning of a power system having objective function and constraints which are bringing in the following sub sections.

7.3.1　Constraints

Let us assume for a power system, the total number of generating unit is n　and　P_{G_j} indicates　the general id power of the jth generator (in MW), P_L indicates the total line loss, P_D indicates the system total load demand, $P_{G_j}^{min}$ and $P_{G_j}^{max}$ are indicating the minimum and maximum operation of generating unit j respectively.

Then, the equality and the inequality constraints for power balance criteria is presented by equation (7.1).

$$\sum_{j=1}^{n} P_{G_j} - P_L - P_D = 0 \qquad\qquad .....(7.1)$$

and

The generator power limitations for each unit as expressed by eq. (7.2).

$$P_{G_j}^{min} \leq P_{G_j} \leq P_{G_j}^{max} \qquad\qquad .....(7.2)$$

7.3.2 The Problem Objective Function

7.3.2.1　The Fuel Cost Function of Power Plant

Let us assume, the total fuel cost $= C_{Total}$, the output power of the jth generating unit $= P_{G_j}$, the cost of the jth generating unit $= C_j$.

Thus, C_{Total} may be expressed as the summation of the all unit fuel costs, which is presented as below:

$$C_{Total} = C_1(P_{G_1}) + C_2(P_{G_2}) + C_3(P_{G_3}) + ... + C_n(P_{G_n}) = \sum_{j=1}^{n} C_j\left(P_{G_j}\right) \qquad .....(7.3)$$

The generator cost curves are characterized by a quadratic function. (second degree polynomial) The total Rs/h fuel cost C_{Total} is stated as:

$$C_{Total} = \sum_{j=1}^{n} \left(a_j P_{G_j}^2 + b_j P_{G_j} + c_j\right) \qquad .....(7.4)$$

where, a_j, b_j, c_j are the cost coefficients.

Analytical calculation for Economic load dispatch without losses:

$$C_T = \sum_i^N C_i\left(P_{G_i}\right) + \lambda\left(\sum_i^N\left(P_{G_i}\right)\right) - \left(P_D\right)$$

where,

$\sum C_i\left(P_{Gi}\right)$ is generation cost for total N units

$\sum\limits_i^N\left(P_{G_i}\right)$ is total generation and (P_D) is total demand.

$$\frac{dc_T}{dp_{G_i}} = \sum_{i=1}^n \frac{dc_1\left(p_{G_i}\right)}{p_{G_i}} - \lambda\left(\sum_{i=1}^n \frac{dp_{G_i}}{dp_{G_i}} - \frac{dp_D}{dp_{G_i}}\right) \qquad .....(7.5)$$

$$\Rightarrow 0 = \sum_{i=1}^n \frac{dc_i\left(p_{G_i}\right)}{p_{Gi}} - \lambda(1 - 0) \qquad .....(7.6)$$

$$\Rightarrow \frac{dc_T}{dp_{G_i}} = \lambda \qquad .....(7.7)$$

$$\Rightarrow \frac{dc_1}{dp_{G_1}} = \frac{dc_2}{dp_{G2}} = \frac{dc_3}{dp_{G3}} = \ldots\ldots\ldots = \frac{dc_n}{dp_{Gn}}. = \lambda \qquad .....(7.8)$$

$\Rightarrow$ Incremental Fuel Cost (IFC) should be same for all the generating units. Therefore, $(IFC)_1 = (IFC)_2 = (IFC)_3 = \ldots\ldots\ldots = (IFG)_N$, where N = Number of generating units

Economic Load Dispatch Calculation with Losses

$$C_T = \sum_i^N C_i\left(P_{G_i}\right) + \lambda\sum_i^N\left(P_{G_i}\right) - \left(P_D\right) - P_L$$

where,

$\sum C_i\left(P_{G_i}\right)$ is generation cost for total N unit,

$\sum\limits_i^N\left(P_{G_i}\right)$ is total generation and (P_D) is total demand and P_L is overall transmission loss.

$$\frac{dc_i}{dp_{G_i}} = \sum_{i-1}^n \frac{dc_1\left(p_{G_i}\right)}{p_{G_i}} - -\lambda\left(\sum_{i=1}^n \frac{dp_{G_i}}{dp_{G_i}} - \frac{dp_D}{dp_{G_i}} - \frac{dp_L}{dp_{Gi}}\right) \qquad .....(7.9)$$

$$\frac{dc_i}{dp_{G_i}} = \lambda\left(1 - \frac{dp_L}{dp_{G_i}}\right) \qquad .....(7.10)$$

$$\lambda = \frac{\frac{dc_i}{dp_{Gi}}}{\left(1 - \frac{dp_L}{dp_{Gi}}\right)} \qquad .....(7.11)$$

$$\lambda = L_1 \frac{dc_1}{dp_{G_1}} = L_2 \frac{dc_2}{dp_{G_2}} = L_3 \frac{dc_3}{dp_{G_3}} = ...L_n \frac{dc_n}{dp_{G_n}} \qquad .....(7.12)$$

$$L_i = \frac{1}{\left(1 - \dfrac{dp_L}{dp_{G_i}}\right)} = \text{penalty factor} \qquad .....(7.13)$$

Problems based on Without Loss

Example 1: The fuel cost of two units are given by:

$$C_1 = C_1(P_{G1}) = 1.0 + 30P_{G1} + 0.4\ P_{G1}^2 \qquad \text{Rs/hr}$$

$$C_2 = C_2(P_{G2}) = 1.5 + 40P_{G2} + 0.4\ P_{G2}^2 \qquad \text{Rs/hr}$$

If the total demand on generators is 200MW, find the economic load scheduling of two units.

Solution:

Neglecting the transmission loss condition for economic load scheduling is given by:

$dC_i/dP_{Gi} = \lambda_i, \lambda = 1, 2, 3....n$

$dC_1/dP_{G1} = 30 + 0.8P_{G1}$ $\qquad\qquad .....(1)$

$dC_2/dP_{G2} = 40 + 0.8P_{G2}$ $\qquad\qquad .....(2)$

For economic load dispatch we have:

$dC_1/dP_{G1} = dC_2/dP_{G2}$

$30 + 0.8P_{G1} = 40 + 0.8P_{G2}$

$0.8P_{G1} - 0.8P_{G2} = 10$ $\qquad\qquad .....(3)$

Also it is given in question that,

$\qquad P_{G1} + P_{G2} = 200\ \text{MW}$ $\qquad\qquad ..... (4)$

On multiplying equation (4) by 0.8 on both sides:

$\qquad 0.8P_{G1} - 0.8P_{G2} = 160$ $\qquad\qquad ..... (5)$

On solving equation (3) & (5)

$\qquad P_{G1} = 106.25\ \text{MW}$

And putting value of P_{G1} in equation (3) we will get the required value of P_{G2}

$\qquad P_{G2} = 93.75\ \text{MW}$

So,

$\qquad P_{G1} = 106.25\ \text{MW}$

$\qquad P_{G2} = 93.75\ \text{MW}$

Example 2: The fuel cost characteristics of two generators are obtained as under:

$$C_1\left(P_{G_1}\right) = 500 + 25P_{G_1} + 0.02P_{G_1}^2 \text{ Rs./hr}$$

$$C_2\left(P_{G_2}\right) = 1500 + 40P_{G_2} + 0.02P_{G_2}^2 \text{ Rs./hr}$$

If the total load supplied is 1000 MW, find the optimal division between two generators.

Solution:

$$C_1\left(P_{G_1}\right) = 500 + 25P_{G_1} + 0.02P_{G_1}^2 \text{ Rs./hr}$$

$$C_2\left(P_{G_2}\right) = 1500 + 40P_{G2} + 0.02P_{G_2}^2 \text{ Rs./hr}$$

The IFC characteristics are:

$$\frac{dc_1}{dp_{G_1}} = 25 + 0.04\,P_{G_1}$$

$$\frac{dc_2}{dp_{G_2}} = 40 + 0.04 P_{G_2}$$

The condition for optimal load division is:

$$\frac{dc_1}{dp_{G_1}} = \frac{dc_2}{dp_{G_2}} = \lambda$$

$$25 + 0.04\,P_{G_1} = 40 + 0.04\,P_{G_2}$$

$$0.04\,P_{G_1} - 0.04\,P_{G_2} = 40 - 25$$

$$0.04\,P_{G_1} - 0.04\,P_{G_2} = 15 \qquad\qquad \text{..... (1)}$$

$$P_{G_1} + P_{G_2} = 1000 \text{ (given)} \qquad\qquad \text{..... (2)}$$

Solving (1) and (2), we get,

$$P_{G_1} = 687.5$$

$$P_{G_2} = 312.5$$

Substituting the values of P_{G_1} and P_{G_2} in $\dfrac{dc_1}{dp_{G_1}}$ or $\dfrac{dc_2}{dp_{G_2}}$ Eq. , we get

$$\frac{dc_1}{dp_{G_1}} = \lambda$$

$$25 + 0.04\,P_{G_1} = 25 + 0.04 \times 687.5 = \lambda$$

$$\lambda = 52.5 \text{ Rs/MWh}$$

The total load of 1000 MW optimally divided between two generators is

$P_{G_1} = 687.5$ MW

$P_{G_2} = 312.5$ MW

 IFC, $\lambda = 52.5$ Rs./MWh

Example 3: The IFCs in rupees per MWh for a plant consisting of two units are

$$\frac{dc_1}{dp_{G_1}} = 0.25 P_{G1} + 50$$

$$\frac{dc_2}{dp_{G_2}} = 0.30\ P_{GL} + 40$$

Calculate the extra cost increased in Rs/hr if a load of 240 MW is scheduled a $P_{G1} = P_{GL} = 120$MW.

Solution: Optimal load condition without losses

$$\frac{dc_1}{dp_{G_1}} = \frac{dc_2}{dp_{G_2}}$$

$0.25\ P_{G1} + 50 = 0.30 P_{G2} + 40$

$0.25 P_{G1} - 0.3\ P_{G2} = -10$ ..... (1)

Given $P_{G1} + P_{G2} = 240$ ..... (2)

Form the Eq.(1) and (2), we get

$P_{G1} = 112.72$ MW & $P_{G2} = 127.27$MW

For the equal distribution of the load power is $P_{G1} = 120$ MW and $P_{G2} = 120$MW.

So the operating cost for the generator 2 will increase because load is increasing from 112.72MW to 120MW.

So increased operating cost for generator unit 2:

$$= \int_{112.72}^{120} \left(0.3 P_{G_2} + 40\right) dP_{G_2}$$

$$= \left[\left(\frac{0.3}{2}\right) P_{G_2}^2 + 40 P_{G2}\right]_{112.72}^{120}$$

$= 0.15(120^2 - 112.72^2) + 40(120 - 112.72)$

$= 543.11$ Rs./hr

The operating cost for the generator will decrease because the load shared is decreasing from 127.27MW to 120MW.

So decreased operating cost for generator unit 1:

$$= \int_{127.27}^{120} (0.25P_{G1} + 50)dP_{G_1}$$

$$= \left[0.125P_{G_1}^2 + 50P_{G_1}\right]_{127.27}^{120}$$

$$= 0.125(120^2 - 127.27^2) + 50(120 - 127.27)$$

$$= -586.57 \text{ Rs./hr}$$

The extra cost is added if both the generator share the equal load

$$= 586.57\text{-}543.11 = 43.46 \text{ Rs./hr}$$

Example 4: The fuel cost curve of two generators are given as:

$$C_1 = 70P_{G1} + 0.12\,P_{G_1}^2 + 1250$$

$$C_2 = 60P_{G2} + 0.010\,P_{G_2}^2 + 1250$$

If total load supplied in 1100 MW, find the optimal dispatch with and without considering the generator limits:

$$70 \text{ MW} \le P_{G1} \le 350 \text{ MW}$$

$$66 \text{ MW} \le P_{G2} \le 1200 \text{ MW}$$

And also comment about the incremental cost of both cases.

Solution:

Given, total load $= P_{G1} + P_{G2} = 1100$ MW

Cost of first unit $C_1 = 70\,P_{G1} + .12\,P_{G1}^2 + 1250$

The IFC of first unit, $\dfrac{dC_1}{dP_{G_1}} = 0.24\,P_{G_1} + 70$ Rs/MWh

Cost of second unit, $C_2 = 60P_{G2} + 0.010\,P_{G_2}^2 + 350$

The IFC cost second unit, $\dfrac{dC_2}{dP_{G_2}} = 0.020P_{G2} + 60$ Rs/MWh

Case 1: without considering generator limits:

For optimal dispatch of load, the necessary condition is

$$\frac{dC_1}{dP_{G_1}} = \frac{dC_2}{dP_{G_2}} = \lambda$$

$$0.24P_{G1} + 70 = 0.020P_{G2} + 60 \hspace{4cm} .....(1)$$

$$0.24P_{G1} - 0.02P_{G2} = -10 \qquad \qquad .....(2)$$

$$0.24P_{G1} + 0.24P_{G2} = 264 \qquad \qquad .....(3)$$

On solving above equations (2) and (3) we get

$P_{G1} = 46.15385$ MW

$P_{G2} = 1053.84$ MW

The above results are for the case without considering the generator limits.

The IFCs are

$$\frac{dC_1}{dP_{G_1}} = 0.24(46.15385) + 70 = 81.07$$

$$\frac{dC_2}{dP_{G_2}} = 0.020(1053.84) + 60 = 81.07$$

Case 2: Considering the generator limits.

$P_{G1} = 46.15385$ MW

$P_{G2} = 1053.84$ MW

It is observed that the real power generation of unit 1 is violating the minimum generation limit. To achieve the optimum operation, fix up the generation of the first unit at its minimum generation i.e.

$P_{G1} = 70$MW. Hence for the load of 1100 MW, $P_{G1} = 70$MW and $P_{G2} = 1030$ MW.

Then, the IFCs are

$$\frac{dC_1}{dP_{G_1}} = 0.24\,(70) + 70 = 86.80 \text{ Rs/MWh}$$

$$\frac{dC_2}{dP_{G_2}} = 0.020(1030) + 60 = 80.60 \text{ Rs/MWh}$$

Hence it is observed that $\dfrac{dC_1}{dP_{G_1}} \neq \dfrac{dC_2}{dP_{G_2}}$, i.e., economic operation is not strictly maintained in this particular condition.

Example 5: A lossless power system has to serve a load of 250MW. There are two generators (G_1 and G_2) in the system with cost curves C_1 and C_2 respectively defined as follows:

$$C_1(P_{G1}) = P_{G1} + 0.055\ P^2_{G1}$$

$$C_2(P_{G2}) = 3P_{G2} + 0.03\ P^2_{G2}$$

where P_{G1} and P_{G2} are the MW injections from generator G_1 and G_2 respectively. Calculate the minimum cost dispatch.

Solution:

$$C_1(P_{G1}) = P_{G1} + 0.055 \times P_{G_1}^2$$

$$dC_1/dP_{G1} = 1 + 0.11P_{G1}$$

$$C_2(P_{G2}) = 3P_{G2} + 0.03 \times P_{G_2}^2$$

$$dC_2/dP_{G2} = 3 + 0.06\,P_{G2}$$

For minimum cost analysis

$$dC_1/dP_{G1} = dC_2/dP_{G2}$$

$$1 + 0.11P_{G1} = 3 + 0.06\,P_{G2}$$

$$0.11P_{G1} - 0.06\,P_{G2} = 2 \qquad\qquad .....(1)$$

$$P_{G1} + P_{G2} = 250 \text{ MW} \qquad\qquad .....(2)$$

$$P_{G1} + P_{G2} = 250 \text{ MW}$$

Solving above equation, (1) & (2), We get

$$P_{G1} = 100 \text{ MW and } P_{G2} = 150 \text{ MW}$$

Example 6: A constant load of 300MW is supplied by two generators 1 and 2, for which the incremental fuel cost is $df_1/dp_1 = 0.10p_1 + 20$ Rs/Mwh; $df_2/dp_2 = 0.12P_2 + 15$ Rs/Mwh with power P in MW and cost F in RS/hr. Determine the most economical division of load.

Solution:

For most economical load division, the incremental fuel rate must be equal

$$0.10P_1 + 20 = 0.12P_2 + 15$$

Also $P_1 + P_2 = 300$

Solving, we get

$$P_1 = 140.9\text{MW and } P_2 = 159.1\text{MW}$$

Example 7: If two plants having cost characteristics as given

$$C_1 = 0.3\,P_{G_1}^2 + 180P_{G1} + 405 \text{ Rs/hr}$$

$$C_2 = 0.45\,P_{G_2}^2 + 120P_{G2} + 300 \text{ Rs/hr}$$

have to meet the following daily load cycle 0 to 24hrs - 210 MW. Find the economic schedule for the given load condition.

Solution:

$C_1 = 0.3\,P_{G1}^2 + 180P_{G1} + 405$

$$\frac{dC_1}{dP_{G_1}} = 0.6P_{G_1} + 180$$

$C_2 = 0.45\,P_{G2}^2 + 120P_{G2} + 300$

$$\frac{dC_2}{dP_{G_2}} = 0.9P_{G2} + 120$$

For 0 to 24hrs: Total load = 210 MW

$P_{G1} + P_{G2} = 210$ MW

The condition for the optimal distribution of load is

$$\frac{dC_1}{dP_{G_1}} = \frac{dC_2}{dP_{G_2}}$$

$0.6P_{G1} + 180 = 0.9_{G2} + 120$

$0.6P_{G1} - 0.9P_{G2} = -60$

By Solving equations, we get

$P_{G2} = 86$ MW

$P_{G1} = 124$ MW

Total operating cost for both the plants for 0 to 24hrs is

$C_1 = 18103.8$ Rs/hr

$C_2 = 22099.2$ Rs/hr

Example 8: The incremental cost characteristics of two thermal plants are given by

$$\frac{dC_1}{dP_{G_1}} = 0.4P_{G1} + 120 \text{ Rs/MWh}$$

$$\frac{dC_2}{dP_{G_2}} = 0.6P_{G2} + 80 \text{ Rs/MWh}$$

Calculate the sharing of a load of 400 MW for most economic operations. If the plants are rated 300 and 500 MW, respectively, what will be the saving in cost in Rs/hr in comparison to the loading in the same proportion to rating.

Solution:

For economic operation,

$$\frac{dC_1}{dP_{G_1}} = \frac{dC_2}{dP_{G_2}}$$

$0.4P_{G1} + 120 = 0.6P_{G2} + 80$

or $0.4P_{G1} - 0.6P_{G2} = -40$

$P_{G1} + P_{G2} = 400$ (given)

Solving above Equation's, we get

▢ $P_{G1} = 200$ MW & $P_{G2} = 200$ MW

Increase in the operation cost for Plant-1 is

$$\int_{200}^{300} \left(0.4P_{G1} + 120\right) dP_{G1} = \left[0.4P_{G1}^2 + 120P_{G1}\right]_{200}^{300}$$

$= 0.2(300^2 - 200^2) + 120 \times 100$

$= 0.2 \times (90000 - 40000) + 12000$

$= 10000 + 12000$

$= 22000$ Rs/hr

Increase in the operation cost for Plant-2 is

$$\int_{200}^{500} \left(0.6P_{G2} + 80\right) dP_{G2} = \left[0.3P_{G2}^2 + 80P_{G2}\right]_{200}^{500}$$

$= 0.3(500^2 - 200^2) + 80 \times 300$

$= 63000 + 24000$

$= 87000$ ₹/hr

∴ Saving in operation cost $= 87000 - 22000 = 65000$ ₹/hr

Example 9: The fuel cost of two generators are given as:

$C_1(P_{G1}) = 50P_{G1} + 0.01\,P_{G1}^2 + 1000$

$C_2(P_{G2}) = 48P_{G2} + 0.003\,P_{G2}^2 + 500$

And if the total load supplied is 700MW, find the optimal dispatch without having transmission loss.

$50\text{MW} \leq P_{G1} \leq 200$ MW

$50\text{MW} \leq P_{G2} \leq 600\text{MW}$

Compare the system's increment at cost with and without generator limit considered.

Solution:

$$\frac{dC_1}{dP_{G_1}} = 50 + 0.02\ P_{G1}$$

$$\frac{dC_2}{dP_{G_2}} = 48 + 0.006\ P_{G2}$$

for economic operation $I_{c(A)} = I_{c(B)} = Y$

$Y = 50 + 0.02\ P_{G1}$

$Y = 48 + 0.006\ P_{G2}$

$P_{G1} + P_{G2} = 700$ MW

Solving these equations, $Y = 46.7$

$P_{G1} = 84.6$ MW

$P_{G2} = 615.4$MW

In the above solution generator limit is not included. If these limits are included it may be seen that generator-2 has violated the limit. Fixing it to the upper most limit then,

$P_{G2} = 600$ MW

And obviously by so that, $P_{G1} = 100$ MW

Therefore $Y_1 = 50 + 0.02 \times 100 = 52$

$Y_2 = 48 + 0.006 \times 600 = 51.6$

Since $\qquad\qquad Y_1 \neq Y_2$

Economic load dispatch is not possible in this condition, incremental cost of generator 1 is more than that of generator 2, and the difference is not much between $Y_1$and Y_2. Hence, the system operation is valid.

Problem based on consideration WITH LOSS

Example 10: A power system has two generators with the following cost curves.

Generator 1: $C_1\ (P_{G1}) = 0.003\ P_{G_1}^2 + 4P_{G1} + 800$ (Thousand Rupees/Hour)

Generator 2: $C_2\ (P_{G2}) = 0.005\ P_{G_2}^2 + 3P_{G2} + 700$ (Thousand Rupees/Hour)

The generator limits are: 50 MW $< P_{G1} <$ 600 MW; 100 MW $< P_{G2} <$ 500 MW

A load demand of 400 MW is supplied by the generator in an optimal manner. Neglecting losses in the transmission network, determine the optimal generation of each generator.

Solution:

$P_{G1} + P_{G2} = 400$ $\qquad\qquad\qquad\qquad\qquad\qquad\qquad\qquad$ .....(1)

$$C_1(P_{G1}) = 0.003\, P_{G_2}^2 + 4P_{G1} + 200$$

Differentiating both sides we get,

$$= 0.006 P_{G1} + 4 \qquad\qquad \text{.....(2)}$$

$$C_2(P_{G2}) = 0.005\, P_{G_2}^2 + 3P_{G2} + 300$$

Differentiating both sides we get,

$$= 0.016 P_{G2} + 3 \qquad\qquad \text{....(3)}$$

Equating Eq. (2) and Eq. (3), we get

$$0.006 P_{G1} - 0.01 P_{G2} = -1 \qquad\qquad \text{.....(4)}$$

Comparing Eq. (1) and (4) we get,

$$P_{G1} = \frac{3}{0.016}$$

$$\Rightarrow P_{G1} = 187.5 \text{ MW}$$

Putting the value of P_{G2} in Eq. (1), we get,

$$P_{G2} = 212.5 \text{ MW}$$

Example 11: A power system consisting of two power plants which are connected by a transmission line is given.

The cost-curve characteristics of the two plants are:

$$C_1 = 0.010\, P_{G_1}^2 + 16P_{G1} + 10 \text{ Rs/hour}$$

$$C_2 = 0.02\, P_{G_1}^2 + 18P_{G2} + 30 \text{ Rs/hour}$$

A loss of 8.212MW is experienced when a power of 100MW is supplied from plant-1 to the load. If the cost of received power is Rs. 20/MWh, find the cost of optimal scheduling of plants and load demand.

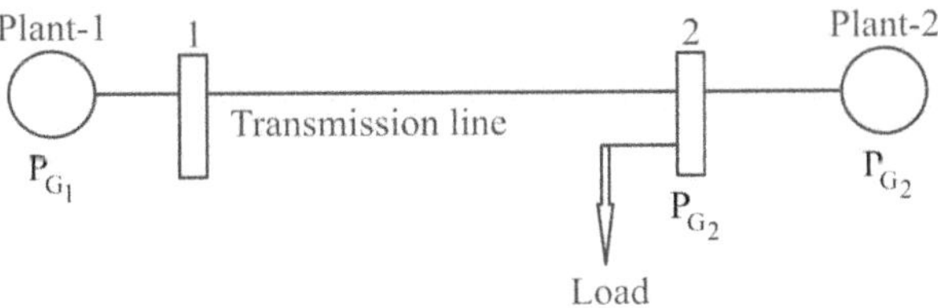

Fig.7.2

Solution:

For two units, $P_L = P_{G1}B_{11}P_{G1} + 2\, P_{G1}B_{12}P_{G2} + P_{G2}B_{21}P_{G1}$

Generator of plant-2 will not affect the losses in the transmission line as the load is located at bus-2 alone.

Therefore, $B_{12} = B_{21} = 0$ and $B_{22} = 0$

$\therefore P_L = P_{G1}^2 B_{11}$

$8.212 = B_{11} \times 100^2$

$B_{11} = 0.0008212 \text{ MW}^{-1}$

$P_L = 0.0008212 \ P_{G1}^2$

Penalty factor method:

The penalty factor of plant-1 is

$$L_1 = \frac{1}{1 - \dfrac{K \partial P_L}{\partial P_G}}$$

$$\frac{\partial P_L}{\partial P_G} = 0.0016424 \ P_{G1}$$

$$L_1 = \frac{1}{1 - 0.0016424 PG1}$$

To obtain optimality the condition is

$$\frac{dC_1}{dP_{G_1}} L_1 = \lambda$$

$$\frac{0.02 PG1 + 16}{1 - 0.0016424 PG1} = 20$$

$P_{G1} = 75.6887 \text{ MW}$

The penalty factor of plant-2 = $L_2 = 1$

$$\frac{dC_2}{dP_{G_2}} = 0.04 P_{G1} + 18$$

To obtain optimality the condition is

$$\frac{dC_2}{dP_{G2}} L_2 = \lambda$$

$(0.04 P_{G2} + 18)1 = 20$

$P_{G2} = 50 \text{MW}$

The transmission loss = PL

$$= P_{G_1}^2 \, B_{11} = 0.0008212 \times (75.6887)^2$$

$$= 4.7044 \text{ MW}$$

$\therefore$ The total load $= P_D = P_{G1} + P_{G2} - P_L$

$= 75.6887 + 50 - 4.7044 = 120.9843 \text{ MW}$

Example 12: The incremental cost of production is given as:

$$\frac{dC_1}{dP_{G_1}} = 0.4\, P_{G1} + 12 \text{ Rs./MWh}$$

Find the incremental cost of power that is received and the penalty factor of the plant.

Solution:

The penalty factor is:

$$L_1 = \frac{1}{1 - \dfrac{\partial P_L}{\partial P_{G_1}}} = \frac{1}{1 - \dfrac{4}{40}} = \frac{10}{9}$$

$\therefore$ Cost of received power

$$= \frac{dC_1}{dP_{G_1}} L_1$$

$$= (0.4 P_{G1} + 12) \times \frac{10}{19}$$

$$= (0.4 \times 40 + 12) \times \frac{10}{9}$$

$$= 31 \text{ Rs/MWh}$$

Example 13: Two-generators are supplying a load of $P_D = 70$ MW, connected at bus. The fuel cost of generators G_1 and G_2 are:

$C_1(P_{G1}) = 8{,}000$ Rs/MWh; $C_2(P_{G2}) = 10{,}000$ Rs/MWh.

And the loss in the line is $P_{\text{loss (pu)}} = 0.25 P^2_{G1 \text{ (pu)}}$, where the loss coefficient is specified in pu on a 100 MVA. Find the economic power generation and loss.

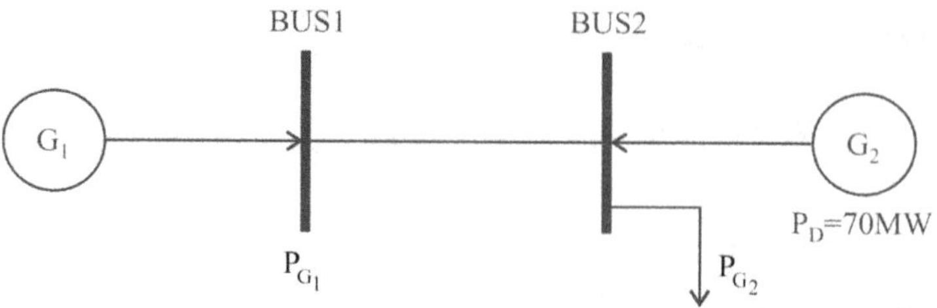

Fig.7.3

Solution:

$$\frac{\partial L}{\partial P_1} = \frac{\partial C_1}{dP_1} + \lambda\left(1 - \frac{\partial P_L}{\partial P_1}\right) \qquad \qquad .....(1)$$

For optimum solution

$$\frac{\partial L}{\partial P_1} = 0$$

From equation (1) we get

$$\frac{\partial C_1}{\partial P_1} + \lambda\left(1 - \frac{\partial P_L}{\partial P_1}\right) = 0$$

Now given, $C_1(P_{G1})$ = 8000 Rs/MWH; $C_2(P_{G2})$ = 10000 Rs/MWh

or, $8000 = \lambda\,(1 - P_1)$ $\qquad \qquad .....(2)$

$$\frac{\partial L}{\partial P_2} = \frac{\partial C_1}{\partial P_2} + \lambda\left(1 - \frac{\partial P_L}{\partial P_2}\right) \qquad \qquad ....(3)$$

or, $10000 = \lambda\,(1 - 0)$ $\qquad \qquad .....(4)$

or, $\lambda = 10000$

Putting the value of λ in eq. (2) we get

$8000 = 10000\,(1 - P_1)$

or, $P_1 = 0.20$ p.u.

or, $P_1 = 0.20 \times 100$ MW

$\qquad = 20$ MW

$P_{loss\,(pu)} = 0.25\,P_{G_1}^2$ (pu)

or, $P_{loss\,(pu)} = 0.25 \times (0.20)^2 = 0.01$ p.u. $= 0.01 \times 100$ MW

$\qquad \qquad = 1$ MW

$P_D + P_{loss} = P_1 + P_2$

or, $P_2 = 50 + 1 - 20$

or, $P_2 = 31$ MW

Example 14: A power system has two generators with the following cost curves.

Generator 1: $C_1(P_{G1}) = 0.003\, P_{G1}^2 + 4P_{G1} + 200$ (Thousand Rupees/Hour)

Generator 2: $C_2(P_{G2}) = 0.004\, P_{G2}^2 + 3P_{G2} + 300$ (Thousand Rupees/Hour)

The generator limits are: 50 MW $< P_{G1} <$ 600 MW; 100 MW $< P_{G2} <$ 500 MW

A load demand of 400 MW is supplied by the generator in an optimal manner. Neglecting losses in the transmission network, determine the optimal generation of each generator.

Solution:

$$P_{G1} + P_{G2} = 400 \qquad\qquad\qquad\qquad\qquad\qquad\qquad .....(1)$$

$$C_1(P_{G1}) = 0.003\, P_{G1}^2 + 4P_{G1} + 200$$

Differentiating both sides we get,

$$= 0.006 P_{G1} + 4 \qquad\qquad\qquad\qquad\qquad\qquad\qquad .....(2)$$

$$C_2(P_{G2}) = 0.004\, P_{G2}^2 + 3P_{G2} + 300$$

Differentiating both sides we get,

$$= 0.008 P_{G2} + 3 \qquad\qquad\qquad\qquad\qquad\qquad\qquad .....(3)$$

Equating Eq. (2) and Eq. (3) we get

$$0.006 P_{G1} - 0.008 P_{G2} = -1 \qquad (4)$$

Comparing Eq. (1) and (4) we get,

$$P_{G2} = \frac{3.4}{0.014}$$

$\Rightarrow P_{G1} = 242.85$ MW

Putting the value of P_{G2} in Eq. (1) we get

$P_{G1} = 157.14$ MW

Example 15: The figure shows a two-generator system supplying a load of $P_D = 40$ MW, connected at bus. The fuel cost of generators G_1 and G_2 are:

$C_1(P_{G1}) = 10,000$ Rs/MWh; $C_2(P_{G2}) = 12,500$ Rs/MWh

And the loss in the line is $P_{loss\,(pu)} = 0.5\, P_{G1(p.u)}^2$, where the loss coefficient is specified in p.u

on a 100 MVA. Find the economic power generation and loss.

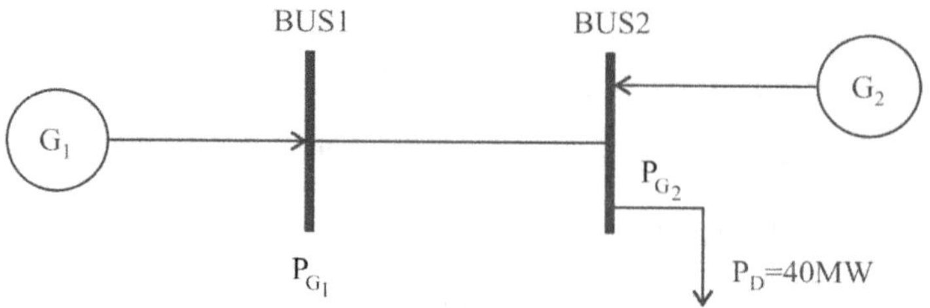

Fig.7.4

Solution:

$$\frac{\partial L}{\partial P_1} = \frac{\partial C_1}{\partial P_1} + \lambda\left(1 - \frac{\partial P_L}{\partial P_1}\right) \qquad\qquad .....(1)$$

For optimum solution,

$$\frac{\partial L}{\partial P_1} = 0$$

From Eq. (1) we get,

$$\frac{\partial C_1}{\partial P_1} + \lambda\left(1 - \frac{\partial P_L}{\partial P_1}\right) = 0$$

Given, $C_1(P_{G1}) = 10{,}000$ Rs/MWh; $C_2(P_{G2}) = 12{,}500$ Rs/MWh

$\Rightarrow 10{,}000 = \lambda\,(1 - P_1)$ $\qquad\qquad\qquad .....(2)$

$$\frac{\partial L}{\partial P_2} = \frac{\partial C_1}{\partial P_2} + \lambda\left(1 - \frac{\partial P_L}{\partial P_2}\right) \qquad\qquad ..... (3)$$

$\Rightarrow 12{,}500 = \lambda\,(1 - 0)$ $\qquad\qquad\qquad ..... (4)$

$\Rightarrow \lambda = 12{,}500$

Putting the value of λ in eq. (2) we get

$10{,}000 = 12{,}500\,(1 - P_1)$

$P_1 = 0.2$ p.u.

$\Rightarrow P_1 = 0.2 \times 100$ MW $= 20$ MW

$P_{loss\,(pu)} = 0.5\ P^2_{G1(p.u)}$

$\Rightarrow P_{loss\,(pu)} = 0.5 \times (0.2)^2 = 0.02$ p.u. $= 0.02 \times 100$ MW $= 2$MW

$P_D + P_{loss} = P_1 + P_2$

$\Rightarrow 40 + 2 - 20 = P_2$

$\Rightarrow P_2 = 22$ MW

Example 16: Two-generators are supplying a load of P_D which is equal to 50 MW, connected at bus. The fuel cost of generators G_1 and G_2 are:

$C_1(P_{G1}) = 4,000$ Rs/MWh; $C_2(P_{G2}) = 5,000$ Rs/MWh

And the loss in the line is $P_{loss\ (pu)} = 0.5\,P^2_{G1(p.u)}$, where the loss coefficient is specified in pu

on a 100 MVA. Find the economic power generation and loss.

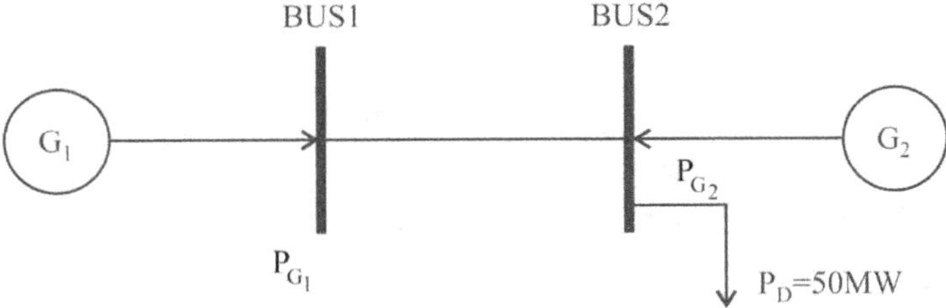

Fig.7.5

Solution:

$$\frac{\partial L}{\partial P_1} = \frac{\partial C_1}{\partial P_1} + \lambda\left(1 - \frac{\partial P_L}{\partial P_1}\right)\qquad\qquad.....(1)$$

For optimum solution,

$$\frac{\partial L}{\partial P_1} = 0$$

From eq. (1) we get,

$$\frac{\partial C_1}{\partial P_1} + \lambda\left(1 - \frac{\partial P_L}{\partial P_1}\right) = 0$$

Given, $C_1(P_{G1}) = 4000$ Rs/MWh

and $C_2(P_{G2}) = 5000$ Rs/MWh

Or, $4000 = \lambda\,(1 - P_1)$ $\qquad\qquad\qquad\qquad\qquad\qquad.....(2)$

$$\frac{\partial L}{\partial P_2} = \frac{\partial C_1}{\partial P_2} + \lambda\left(1 - \frac{\partial P_L}{\partial P_2}\right)\qquad\qquad....(3)$$

Or, $5000 = \lambda\,(1 - 0)$ $\qquad\qquad\qquad\qquad\qquad\qquad....(4)$

Or, $\lambda = 5000$

Putting the value of λ in eq. (2) we get

$4000 = 5000\,(1 - P_1)$

Or, $P_1 = 0.20$ p.u.

Or, $P_1 = 0.20 \times 100$ MW

$= 20$ MW

$$P_{loss\,(pu)} = 0.5\,P^2_{G1(p.u.)}$$

$$\text{Or, } P_{loss\,(p.u)} = 0.5 \times (0.20)^2 = 0.02 \text{ p.u.} = 0.02 \times 100 \text{ MW}$$

$$= 2 \text{ MW}$$

$$P_D + P_{loss} = P_1 + P_2$$

$$\text{Or, } P_2 = 50 + 2 - 25$$

$$= 27 \text{ MW}$$

Example 17: The figure shows a two-generator system supplying a load of $P_D = 60$ MW, connected at bus. The fuel cost of generators G_1 and G_2 are:

$$C_1(P_{G1}) = 5,000 \text{ Rs/MWh}; \quad C_2(P_{G2}) = 7,500 \text{ Rs/MWh}$$

And the loss in the line is $P_{loss\,(p.u)} = 0.5\,P^2_{G1(p.u)}$, where the loss coefficient is specified in pu on a 100 MVA .Find the economic power generation and loss.

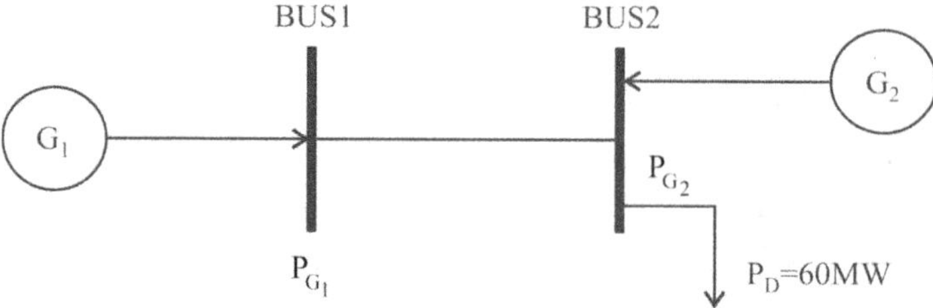

Fig.7.6

Solution:

$$\frac{\partial L}{\partial P_1} = \frac{\partial C_1}{\partial P_1} + \lambda\left(1 - \frac{\partial P_L}{\partial P_1}\right) \qquad \text{....(1)}$$

For optimum solution,

$$\frac{\partial L}{\partial P_1} = 0$$

From eq.(1) we get,

$$\frac{\partial C_1}{\partial P_1} + \lambda\left(1 - \frac{\partial P_L}{\partial P_1}\right) = 0$$

Given, $C_1(P_{G1}) = 5,000 \text{ Rs/MWh}; \quad C_2(P_{G2}) = 7,500 \text{ Rs/MWh}$

$$\Rightarrow 5,000 = \lambda\,(1 - P_1) \qquad \text{.....(2)}$$

$$\frac{\partial L}{\partial P_2} = \frac{\partial C_1}{\partial P_2} + \lambda\left(1 - \frac{\partial P_L}{\partial P_2}\right) \qquad \text{.....(3)}$$

$$\Rightarrow 7,500 = \lambda(1 - 0) \qquad \text{.....(4)}$$

$\Rightarrow \lambda = 7,500$

Putting the value of λ in eq.(2) we get

$5,000 = 7,500 \, (1 - P_1)$

$\Rightarrow P_1 = 0.33$ p.u.

$\Rightarrow P_1 = 0.33 \times 100$ MW $= 33$ MW

$P_{loss \, (pu)} = 0.5 \, P^2_{G1(p.u)}$

$\Rightarrow P_{loss \, (pu)} = 0.5 \times (0.33)^2 = 0.054$ p.u. $= 0.054 \times 100$ MW $= 5.445$MW

$P_D + P_{loss} = P_1 + P_2$

$\Rightarrow 60 + 5.445 - 33 = P_2$

$\Rightarrow P_2 = 32.445$ MW

Example 18: Incremental fuel costs for a power plant consisting of three generating units are:

$IC_1 = 10 + 0.2P_1, \; IC_2 = 20 + 0.4 \, P_2, \; IC_3 = 30$

where P_i is the power in MW generated by unit I, for i $= 1, 2$ and 3.

Assume that all the three units are operating all the time. Minimum and maximum loads on each unit are 40MW and 200MW respectively. If the plant is operating on economic load dispatch to supply the total power demand of 500MW, find the power generated by each unit.

Solution:

When P_1 is minimum i.e. $P_1 = 40$ MW

$IC_1 = 10 + 0.2P_1$

$\quad = 10 + 0.2(40)$

$\quad = 18$ MW

When P_2 is minimum i.e. $P_2 = 40$ MW

$IC_2 = 20 + 0.4P_2$

$\quad = 20 + 0.4(40)$

$\quad = 36$ MW

For minimum value of P_1 and P_2

IC_1 and $IC_2 > IC_3$

Therefore

$\quad P_3 = 200$MW

Remaining load is shared by unit 1 and 2.

So $P_1 + P_2 = 500 - 200 = 300$ MW $\qquad\qquad\qquad\qquad\qquad$ .....(1)

For optimal operation

$IC_1 = IC_2$

$10 + 0.2P_1 = 20 + 0.4P_2$

$0.2P_1 - 0.4P_2 = 10$ $\qquad\qquad\qquad\qquad\qquad\qquad\qquad$ .....(2)

Solving eq. (1) and (2), we get

$P_1 = 216.667$ MW

$P_2 = 83.333$ MW

Therefore powers generated by each unit are

$P_1 = 216.667$MW

$P_2 = 83.333$ MW

$P_3 = 200$MW

Example 19: In a power system, the fuel inputs per hour of plants 1 and 2 are given as

$F_1 = 0.20P_1^2 + 30\,P_1 + 100$ Rs. Per hour

$F_2 = 0.25P_2^2 + 40P_2 + 80$ Rs. Per hour

The limits of generators are $20 \leq P_1 \leq 80$ MW; $40 \leq P_2 \leq 200$ MW. Find the economic operating schedule of generation, if the load demand is 130 MW, neglecting transmission losses.

Solution:

$P_1 + P_2 = 130$ $\qquad\qquad\qquad\qquad\qquad\qquad\qquad\qquad$ .....(1)

$F_1 = 0.20\,P_1^2 + 30\,P_1 + 100$

Differentiating on both sides we get

$\qquad = 0.4\,P_1 + 30$ $\qquad\qquad\qquad\qquad\qquad\qquad\qquad$ .....(2)

$F_2 = 0.25\,P_2^2 + 40P_2 + 80$

Differentiating on both sides we get

$\qquad = 0.5\,P_2 + 40$ $\qquad\qquad\qquad\qquad\qquad\qquad\qquad$ .....(3)

Equating Eq. (2) and Eq. (3) we get,

$0.4\,P_1 + 30 = 0.5\,P_2 + 40$

$0.4\,P_1 - 0.5\,P_2 = 10$

Multiplying both sides by 2 we get,

$0.8\,P_1 - P_2 = 20$ $\qquad\qquad\qquad\qquad\qquad\qquad\qquad\qquad$ (4)

Equating Eq. (1) and Eq. (4) we get,

$1.8\,P_1 = 150$

$P_1 = 83.33$

But as the problem says that P_1 must vary between $20 \leq P_1 \leq 80$, so we will take

$P_1 = 80$

By putting the value of P_1 in eq. (1) we get

$P_2 = 50$

Economic Scheduling: Economic scheduling gives the allocation of the load among the various units in service in such a way that total cost of generation is minimum. Unit commitment problem decides which units are kept ON and which units are kept OFF to supply a given load demand for particular period of time.

Example 20: Incremental cost characteristics of two generators are delivering 300 MW as:

$dF_1/dP_1 = 30 + 0.15\,P_1$

$dF_2/dP_2 = 24 + 0.3\,P_2$

Determine the economic scheduling for two generators.

Solution:

$$P_1 + P_2 = 300 \qquad\qquad\qquad ....(1)$$

For economic operation,

$$dF_1/dP_1 = dF_2/dP_2$$

or, $30 + 0.15\,P_1 = 24 + 0.3\,P_2$

or, $0.15\,P_1 - 0.3\,P_2 = -6$

or, $1.5\,P_1 - 3\,P_2 = -60 \qquad\qquad\qquad .....(2)$

by solving (1) & (2) we get,

$P_1 = 186.667$ MW

$P_2 = 113.33$ MW

Example 21: A power system has two generators with the following cost curves.

Generator 1: $C_1\,(P_{G1}) = 0.006\,P_{G_1}^2 + 8P_{G1} + 350$ (Thousand Rupees/Hour)

Generator 2: $C_2\,(P_{G2}) = 0.006\,P_{G_2}^2 + 7P_{G2} + 400$ (Thousand Rupees/Hour)

The generator limits are: 100 MW $< P_{G1} < 650$ MW; 50 MW $< P_{G2} < 500$ MW

A load demand of 600 MW is supplied by the generator in an optimal manner. Neglecting losses in the transmission network, determine the optimal generation of each generator.

Solution:

Given,

$$P_{G1} + P_{G2} = 600 \qquad \qquad \dots(1)$$

$$C_1(P_{G1}) = 0.006\,P_{G2}^2 + 8P_{G1} + 350$$

Differentiating both sides we get,

$$= 0.012P_{G1} + 8 \qquad \qquad \dots(2)$$

$$C_2(P_{G2}) = 0.006P_{G2}^2 + 7P_{G2} + 400$$

Differentiating both sides we get,

$$= 0.012P_{G2} + 7 \qquad \qquad \dots(3)$$

Equating Eq. (2) and Eq. (3) we get

$$0.012P_{G1} - 0.012P_{G2} = -1$$

$$\Rightarrow \qquad \qquad P_{G1} - P_{G2} = \frac{1}{0.012} \qquad \qquad \dots(4)$$

Comparing Eq. (1) and (4) we get,

$$2P_{G1} = 600 - \frac{1}{0.012}$$

$$\Rightarrow P_{G1} = 258.33 \text{ MW}$$

Putting the value of P_{G1} in Eq. (1) we get

$$P_{G2} = 341.66 \text{ MW}$$

Example 22: There are two power plants connected through a transmission line at plant 2 as shown below. 100 MW is transmitted from plant 1 resulting in 10 MW transmission loss.

$$C_1 = 0.05\,P_{G1}^2 + 15\,P_{G1} \text{ Rs/hr}$$

$$C_2 = 0.06\,P_{G2}^2 + 11\,P_{G2} \text{ Rs/hr}$$

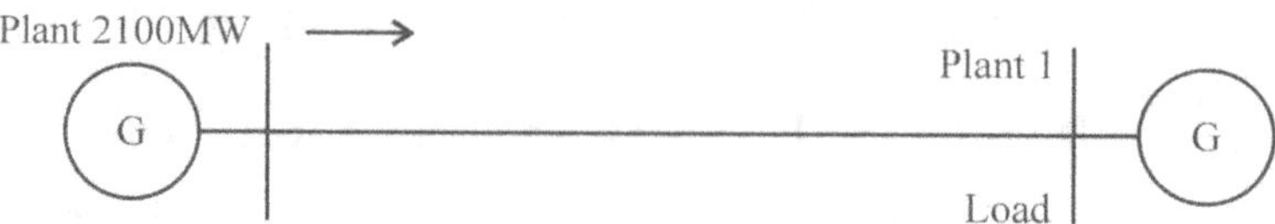

Fig.7.7

Find the optimum generation for $\lambda = 23$

Solution: $P_L = 10$ MW

$$P_L = B_{11} P_{G1}^2 + B_{22} P_{G2}^2 + P_{G_1} B_{12} P_{G_2} + P_{G_2} B_{21} P_{G_1} \qquad [B_{22}=0, \qquad B_{21}=0, B_{12}=0]$$

$$P_L = B_{11} P_{G1}^2 \qquad B_{11} = \frac{PL}{PG_1} = \frac{10}{100 \times 100} = 10^{-3} \left[\lambda = \frac{\frac{\partial C_1}{\partial PG1}}{1 - \frac{\partial PL}{\partial PG_2}} = \frac{\frac{\partial C_2}{\partial PG2}}{1 - \frac{\partial PL}{\partial PG_2}} \right]$$

$$\lambda = L_1 \frac{\partial C_1}{\partial PG_2} = L_2 \frac{\partial C_2}{\partial PG_2} \left[L_1 = \frac{1}{\frac{\partial PL}{\partial PG_1}} = \frac{1}{1 - \frac{\partial PL}{\partial PG_1}} \right]$$

So for generator 1,

$$\lambda = L_1 \frac{\partial C_1}{\partial PG_1}$$

$$P_L = B_{11} P_{G2}^2$$

$$P_L = 10^{-3} P_{G1}^2$$

$$\frac{\partial PL}{\partial PG_1} = 2 \times 10^{-3} P_{G1}$$

$$L_1 = \frac{1}{1 - 0.02 PG1}$$

$$23 = \frac{1 \times (0.1 PG1 + 15)}{(1 - 0.002 PG1)}$$

$$23 - 0.46\, P_{G1} = 0.1 P_{G1} + 15$$

$$8 = 0.146\, P_{G1}$$

$$P_{G1} = 54.79 \text{ MW}$$

For generator 2,

$$L_2 = 1 \quad \text{Since,} \ \frac{\partial P_L}{\partial P_{G_2}} = 0$$

$$\lambda = \frac{\partial C_2}{\partial P_{G_2}}$$

$$23 = (0.12\, P_{G2} + 11)$$

$$P_{G2} = \frac{12}{0.12} = 100 \text{ MW}$$

Example 23: Determine the incremental cost of received power and the penalty factor of the plant as shown in Fig.(7.8), if the incremental cost of production is.

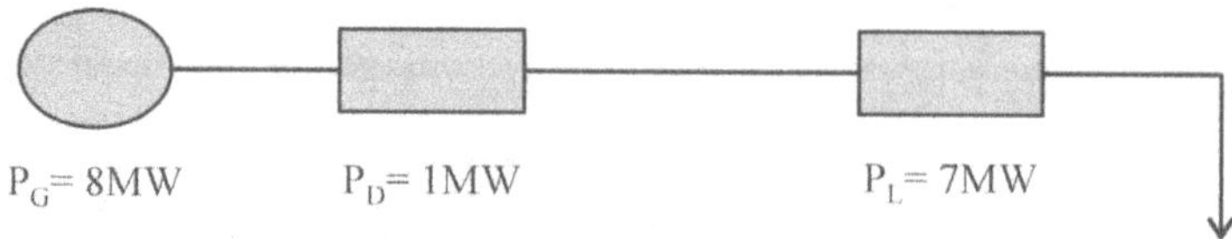

Fig.7.8

$$\frac{dC_1}{dP_{G_1}} = 0.2P_{G1} + 4.0 \text{ Rs/MWh}$$

Solution:

The penalty factor $= L_1 = \dfrac{1}{1 - \frac{\partial PL}{\partial PG_1}} = \dfrac{8}{7}$

Therefore, cost of received power $= \dfrac{\partial C_1}{\partial PG_1} L_1$

$$= (0.2\ P_{G1} + 4) \times \frac{8}{7}$$

$$= (0.2 \times 8 + 4) \times \frac{8}{7}$$

$$= 5.6 \times \frac{8}{7}$$

$$= 6.4 \text{ Rs/MWh}$$

Example 24: The IFC for two plants are:

$$\frac{\partial C_1}{\partial PG_1} = 0.05\ P_{G1} + 15, \quad \frac{\partial C_2}{\partial PG_2} = 0.06\ P_{G2} + 14$$

The loss coefficient are given as, $B_{11} = 0.013/\text{MW}$, $B_{12} = -0.004/\text{MW}$, $B_{22} = 0.0030/\text{MW}$, $\alpha = 25$ Rs/MWh. Find the real power generation, total load demand, and the transmission power loss.

Solution:

$$P_{G1} + P_{G2} = P_D = P_L$$

And transmission loss, $P_L = \sum_{p=1}^{n} P_{GP} B_{pq} P_{Gq}$

For number of plants, n = 2, we have

$$= P_{G1} B_{11} P_{G1} + P_{G1} B_{12} P_{G2} + P_{G2} B_{21} P_{G1} + P_{G2} B_{22} P_{G2}$$

$$= 0.013\ P_{G_1}^2 + 2(-0.004)\ P_{G1}\ P_{G2} + 0.0030\ P_{G2}^2$$

$$= 0.013\ P_{G_1}^2 - 0.008\ P_{G1}\ P_{G2} + 0.0030\ P_{G2}^2$$

The ITL and Penalty factor of Plant-1 is,

$$(ITL)_1 = \frac{\partial P_L}{\partial P_{G_1}} = 2(0.0013)\ P_{G1} - 0.008\ P_{G2}$$

Penalty factor of Plant 1:

$$L_1 = \frac{1}{1 - \frac{\partial P_L}{\partial G_1}} = \frac{1}{1 - \left(0.026PG1 - 0.08PG2\right)}$$

The ITL of plant 2:

$$(ITL)_2 = \frac{\partial P_L}{\partial P_{G_2}} = 0.006\ PG_2 - 0.008\ PG_1$$

And penalty factor of Plant-2

$$L_2 = \frac{1}{1 - \frac{\partial P_L}{\partial G_2}} = \frac{1}{1 - \left(0.006 P_{G_2} - 0.008 P_{G_1}\right)}$$

Condition for optimum operation is

$$\frac{\partial C_1}{\partial PG_1} L_1 = \frac{\partial C_2}{\partial PG_2} L_2 = \alpha$$

and
$$\frac{\partial C_1}{\partial PG1} L_1 = \alpha$$

$$0.005 P_{G1} + 0.65 P_{G1} - 0.2 P_{G2} = 10$$

$$0.7\ P_{G1} - 0.2 P_{G2} = 10$$

$$\frac{\partial C_2}{\partial PG_2} L_2 = \alpha$$

$$0.21 P_{G2} - 0.2 P_{G1} = 11$$

Solving above equation,

We get,

$$P_{G1} = 40.18\ MW$$

$$P_{G2} = 90.65\ MW$$

Transmission loss, $P_L = 0.013(40.18)^2 - 0.008 \times (40.18) \times (90.65) + (90.65)^2 0.0030$

$= 16.50135$ MW

Total load, $P_D = P_{G1} + P_{G2} - P_L$

$\qquad = 40.18 + 90.65 - 16.50$

$\qquad = 114.33$ MW

Example 26: On a system consisting of two generating plants, the incremental costs in Rs/MWh with P_{G1} and P_{G2} in MW are

$$\frac{\partial C_1}{\partial PG_1} = 0.006P_{G1} + 6.0, \quad \frac{\partial C_2}{\partial PG_2} = 0.01P_{G2} + 7.0$$

The system is operating on economics dispatch with $P_{G1} = P_{G2} = 300$ MW and $\dfrac{\partial PL}{\partial PG_2} = 0.2$.

Find the penalty factor of Plant-1.

Solution: Given that the system operates on economic dispatch with $P_{G1} = P_{G2} = 300$MW, the condition for this optimal operation when considering the transmission loss is

$$\frac{\partial C_1}{\partial PG_1} L_1 = \frac{\partial C_2}{\partial PG_2} L_2 = \alpha$$

And also given the ITL of Plant 2,

$$(ITL)_2 = \frac{\partial P_L}{\partial G_2} = 0.2$$

The penalty factor of Plant -2,

$$L_2 = \frac{1}{1 - \frac{\partial PL}{\partial PG2}} - \frac{1}{0.8} = 1.25$$

Therefore, For optimal condition,

$$\frac{\partial C_1}{\partial PG_1} L_1 = \frac{\partial C_2}{\partial PG_2} L_2$$

$(0.006P_{G1} + 6) L_1 = (0.01P_{G2} + 7)L_2$

Or $7.8L_1 = 12.5$

Or $L_1 = 1.6025$

Therefore, Penalty factor of Plant-1 = 1.6025

Exercise

Multiple Choice Questions

1. The fuel cost function in rupees / hour for two 750MW thermal power plants are given below

   Plant 1: $C_1 = 250 + 8P_1 + 0.006P_1^2$

   Plant 2: $C_2 = 400 + aP_2 + 0.005P_2^2$

   Where P_1 and P_2 are power generated by plant 1 and plant 2 respectively in MW and a is constant. The incremental cost of power (λ) is 10 rupees per MWh. The two thermal power plants together meet the total power demand of 600 MW. The optimal generation plant 1 and plant 2 in MW, respectively are

   a) 235, 265   b) 166, 434   c) 325, 275   d) 350, 250

2. A measure of the penalty factor in the economical operation of power systems is

   a) Fuel cost   b) generation cost   c) Line loss   d) power delivered

3. Maximum demand on generating stations divided by installed capacity of plant results in:

   a) load factor   b) operating factor   c) utilization factor   d) capacity factor

4. The following factors affect electrical supply costs and tariffs:

   a) (i) standing charges that are independent of the output and (ii) running or operating charges that are proportional to the output.

   b) (i) standing charges that are proportional of the output and (ii) running or operating charges that are independent of the output.

   c) (i) standing charges that are proportional to the output and (ii) running or operating charges that are proportional to the output.

   d) (i) standing charges that are independent of the output and (ii) running or operating charges that are independent to the output.

5. What is a plant's penalty percentage if its incremental loss is 0.75?

   a) 5   b) 4   c) 4.2   d) 1.25

6. Increased of penalty factors between a producing plant's current and system load current have the following effects on a power system's economic operation:

   a) hold the plant load constant   b) increase load on plant

   c) decrease load on the generating plant.   d) increase power factor.

7. Which of the following expenses is a power system's operational cost?

   a) fuel cost   b) insurance

   c) initial cost   d) taxes and interest

8. The ratio between sum of maximum demand of separate consumers to maximum demand of the station is

    a) load factor   b) utility factor   c) diversity factor   d) peak factor

9. Load factor =

    a) ratio between maximum peak load to average load

    b) ratio between total load to maximum load

    c) ratio between average load to maximum load

    d) ratio between minimum load to total load

10. The ratio between energy produced by a power plant to the installed capacity of the plant is

    a) Plant load factor   b) demand factor

    c) average load factor   d) plant use factor

Answers

1.(b)   2.(c)   3.(c)   4.(a)   5.(b)   6.(c)   7.(a)   8.(c)   9.(c)   10(d)

1. The fuel cost characteristics of two generators are obtained as under:

$$C_1(PG_1) = 600 + 20\,P_{G1} + 0.06\,P_{G_1}^2 \ \text{Rs./hr}$$

$$C_2(PG_2) = 1200 + 60\,P_{G2} + 0.02\,P_{G_2}^2 \ \text{Rs./hr}$$

   If the total load supplied is 800 MW, find the optimal division between two generators.

   The fuel cost curve of two generators are given as:

$$C_1 = 60P_{G1} + 0.15\,P_{G1}^2 + 1500$$

$$C_2 = 40P_{G2} + 0.020\,P_{G2}^2 + 700$$

   If total load supplied in 1200 MW, find the optimal dispatch with and without considering the generator limits:

$$100 \text{ MW} \leq P_{G1} \leq 150 \text{ MW}$$

$$200 \text{ MW} \leq P_{G2} \leq 1200 \text{ MW}$$

   And also comment about the incremental cost of both cases.

2. A constant load of 800 MW is supplied by two generators 1 and 2, for which the incremental fuel cost is $df_1/dp_1 = 0.20p_1 + 40\text{Rs/Mwh}$; $df_2/dp_2 = 0.24p_2 + 25\text{Rs/Mwh}$ with power P in MW and cost F in RS/hr. Determine the most economical division of load.

3. Power system has two generators with the following cost curves.

   Generator 1: $C_1(P_{G1}) = 0.004P^2_{G1} + 6P_{G1} + 300$ (Thousand Rupees/Hour)

   Generator 2: $C_2(P_{G2}) = 0.006P^2_{G2} + 4P_{G2} + 400$ (Thousand Rupees/Hour)

   The generator limits are: 100 MW $< P_{G1} <$ 400 MW; 100 MW $< P_{G2} <$ 600 MW

   A load demand of 500 MW is supplied by the generator in an optimal manner. Neglecting losses in the transmission network, determine the optimal generation of each generator.

4. There are two power plants connected through a transmission line at plant 2 as shown below. 100 MW is transmitted from plant 1 resulting in 10 MW transmission loss.

   $C_1 = 0.05\ P^2_{G_1} + 15\ P_{G1}$ Rs/hr

   $C_2 = 0.06\ P^2_{G_2} + 11\ P_{G2}$ Rs/hr

   Find the optimum generation for $\lambda = 23$

5. The IFC for two plants are:

   $dC_1/dP_{G1} = 0.055P_{G1} + 18$ Rs/MWh

   $dC2/dPG2 = 0.08PG1 + 16$ Rs/MWh

   The loss coefficients are given as: $B_{11} = 0.0016$/MW, $B_{12} = -0.0008$/MW, $B_{22} = 0.0024$/MW. For $\lambda = 10$ Rs/MWh. Find the real power generations, total load demand and the transmission Power loss.

6. The IFC for two plants are:

$$\frac{dC_1}{dP_{G_1}} = 0.06\ P_{G1} + 15, \quad \frac{dC_2}{dP_{G_2}} = 0.08\ P_{G2} + 14$$

   The loss coefficient are given as, $B_{11} = 0.014$/MW, $B_{12} = -0.006$/MW, $B_{22} = 0.0020$/MW, $\alpha = 20$ Rs/MWh. Find the real power generation, total load demand, and the transmission power loss.

Chapter 8

Power Angle Stability and its Analysis

8.1. Introduction

The different components of Power system are synchronous generators, transformers, switchgear, transmission lines, distribution lines and different types of loads etc. Any fault or any disturbance like change of input mechanical power in terms of steam flow to the turbine, change of electrical loads, line outage, and generator outage may change the operating state which lead to failure of synchronism of power system components. This will affect on the other components of power system. This phenomenon is referred as rotor angle stability. Rotor angle stability is apprehension with the capability of interconnected synchronous machines of a power system to stay in synchronism under normal operating conditions after subsequent to a disturbance. During normal condition, the input mechanical torque and the output electromagnetic torque of each generator is equal, and thus speed of rotor of generator unvarying. But due to mismatch of the input mechanical torque and the output electromagnetic torque of any generator, the equilibrium operating state of power system may be disturbed. As a result in stability may take place in the appearance of rising or declining angular swings of some generators which lead to their loss of synchronism with other generators.

8.2 Definitions

Stability

Stability is referring a state in which different synchronous machines in the power system stay in synchronism.

Instability

Instability is referring a state in which different synchronous machines in the power system stay in involving loss of synchronism.

8.3 Types of Stability

Power system stability problems are classified into two basic types; namely

(i) Steady state:

The ability of an electrical machine or power system to restore its original state after disturbance or we can suggest that the angle of the rotor angle shifts in slight disturbance.

(ii) Transient state stability:

It is characterized as the power system's ability to return after a major disruption to its normal conditions. Because of the sudden removal of load, line switching operations, th e major disturbance occurs in the system; fault occurs in the system, a line's sudden out age, etc.

Steady State Stability Limit

It denotes to the maximum power flow through a given point without causing a loss of equilibrium when the power is increased very gradually.

8.4 Power Angle Equation

Let us start with a very easy power network as shown in Fig.(8.1) which comprises of a synchronous generator delivering power to synchronous motor through a transmission line.

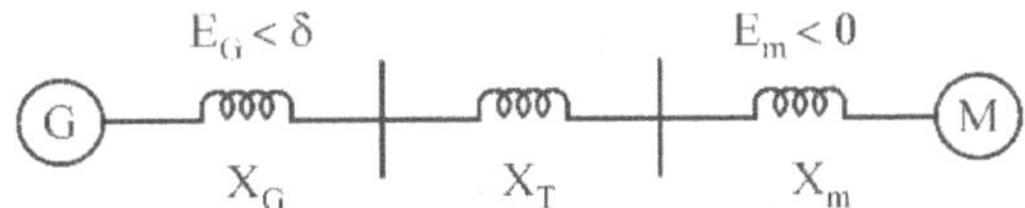

Fig.8.1 Two Machine System.

A synchronous machine may be represented by a constant voltage source in series with a reactance X. The reactance X, is expressed based upon the state under study. If it is analyzed during transient state of the machine then the reactance X may be presented as sub transient reactance X'_d. If it is analyzed during transient state of the machine then the reactance X may be presented as transient reactance X'_d. If it is analyzed during steady state of the machine then the reactance X may be presented as steady state synchronous reactance X_d. Thus the above one line illustration may be drawn in terms of reactance diagram as shown in figure.

Where the generator voltage is symbolized by E_G in series with its own reactance which is characterized by X_G.

The motor voltage is symbolized by E_M in series with its own reactance which is characterized X_M.

The transmission line with leakage reactance is symbolized by X_L.

Whole reactance between the two machines is represented by

$$X = X_G + X_T + X_M$$

From the Fig 8.1, the relation between $\overline{E}_G$ and $\overline{E}_M$ may be written as, $\overline{E}_G = \overline{E}_M + jX\overline{I}$.

Thus the current is, $\overline{I} = \dfrac{\overline{E}_G - \overline{E}_M}{jX}$

The resistances of the machines and the transmission line are less compared to reactance. Hence, resistances may be neglected.

The active power delivered by the generator is the input power of the motor. This active power 'P', may be presented as

$P = $ Real part of $\left(\overline{E}_G \times \overline{I}\right)$

$$= \mathrm{Re}\left[\overline{E}_G\left(\frac{\overline{E}_G - \overline{E}_M}{jX}\right)\right]$$

Let $\overline{E}_M = E_M < 0$, $\overline{E}_G = E_G < \delta$,

$\therefore$ $\overline{E}_G = EG < -\delta$

Now putting the above, we get

$$P = \mathrm{Re}\left[E_G < -\delta\left(\frac{E_G < \delta - E_M < 0}{X < 90^\circ}\right)\right]$$

$$= \mathrm{Re}\left[\frac{E_G^2}{X} < -90^\circ - \frac{E_G E_M}{X} < -90 - \delta\right]$$

$\therefore$ Real power P is given by

$$P = -\frac{E_G E_M}{X}\cos\left(-90 - \delta\right)$$

$$= -\frac{E_G E_M}{X}\cos\left(90 + \delta\right)$$

$$P = \frac{E_G E_M}{X}\sin\delta \qquad\qquad .....(8.1)$$

The above equation demonstrates that the power delivered from the generator to the motor fluctuate with the sine of the displacement angle δ between the two rotors. This equation is identified as the power angle equation. From the curve it may be observed that maximum power transfers when $\sin\delta = 1$ i.e., $\delta = 90$. Thus, $P_{max} = \dfrac{E_G E_M}{X}$. This is called, the steady state stability limit. An important information regarding stability may be obtained from the slope power angle curve. If the slope $\dfrac{dP}{d\delta}$ is positive $(-90^0 \le \delta \le 90^\circ)$, it indicates that raise of displacement angle which lead to increase transmitted power and consequently the system will be stable. The value negative slope point outs that the system is unstable.

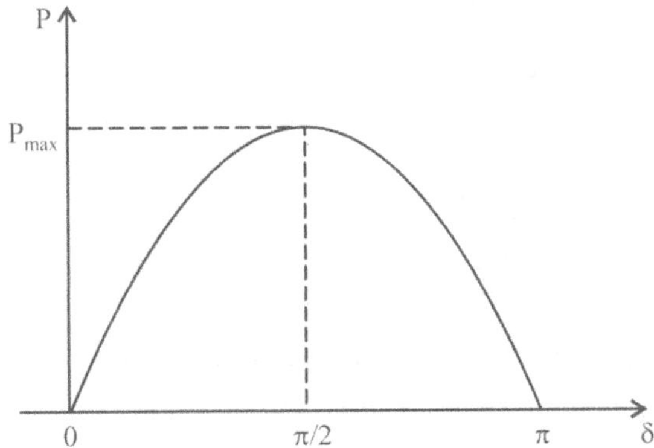

Fig.8.2 Power Angle Curve.

When the slope $\dfrac{dP}{d\delta}$ is positive ($-90^0 \le \delta \le 90^0$), it means that an increase in displacement angle results in an increase in transmitted power and hence the system will be stable.

If $\dfrac{dP}{d\delta}$ is negative, it indicates that the system is unstable.

Solved Problem

1. Two synchronous machines of equal rating have internal voltage of 1.1 + j0.5 and 0.8 − j0.4 per unit voltages respectively. The machines are connected by a line of 50 km length having only reactance and the second machine receives power of 0.9 per unit. Determine the reactance of the line per km length. Assume that there is no internal reactance for simplification.

Solution:

Given

$$\overline{E}_G = 1.1 + j0.5 = 1.21 < 24.4^0$$

$$\overline{E}_M = 0.8 - j0.4 = 0.89 < -26.6^0$$

P = 0.9, length of transmission line = 50 km

$$X_G = X_M = 0$$

$$\delta = 24.4^0 - (-26.6^0)$$

$$= 51^0$$

We know that, the power angle equation is,

$$P = \frac{E_G E_M}{X} \sin \delta$$

$$0.9 = \frac{1.21 \times 0.89}{X} \sin 51^\circ$$

$$X = \frac{0.8369}{0.9}$$

$$= 0.9299 \text{ p.u.}$$

Since $X_G = X_M = 0$, X denotes the reactance of the transmission line.

X = 0.9299 p.u.

$$X \text{ per km} = \frac{0.9299}{50}$$

$$= 0.0186 \text{ p.u}$$

8.5 Techniques for getting better Steady State Stability

The Steady state stability limit may be presented by section 8.1

$$P_{max} = \frac{E_G E_M}{X}$$

From the equation it has been observed that P_{max} may be increased by

(i) increasing any or in cooperation E_G and E_M and (ii) reducing X.

8.6 Swing Equation

In a synchronous generator, the input power by means of mechanical or shaft torque used to apply and the output power is obtained in the form of electromagnetic torque and finally from the generator terminal as electrical power and for synchronous motor it is vice versa. A power system comprises of many synchronous machines which runs synchronously for all operating circumstances. Generally at steady stable state, the rotor axis place and magnetic field axis are unchanging. The angle between the rotor axis and magnetic field axis is identified as power angle. The rotor will decelerate or accelerate for the disturbances as explained in section 8.1. The change of load and with respect to time due to the disturbance may be expressed as a non-linear 2nd order differential equation and referred as swing equation which provide a brief idea of rotor angle stability of synchronous machine.

8.7 Analytical Concept of Swing Equation

Swing equation is the action of the synchronous machine during transient condition. Let θ be the angular position of the motor at any instant t. θ continuously changes with time. It is easy to compute θ with respect to a reference axis that is revolving at synchronous speed. Here, δ is the angular displacement of the rotor in electrical degrees from the synchronously rotating reference axis and ω_s is the synchronous speed in electrical radians, then θ may be presented as the sum of two quantities. The first part is the time varying angle $\omega_s t$ on the rotating reference axis and the second part is the torque angle δ of the rotor with respect to the rotating reference axis. It can be presented by the eq.(8.2).

$$\theta = w_s t + \delta \text{ electrical radians} \qquad \qquad ....(8.2)$$

Differentiating equation (8.2) with respect t we get

$$d\theta/dt = ws + d\delta/dt \qquad \qquad .....(8.3)$$

Differentiating equation (8.3) give

$$d^2\theta/dt^2 = d^2\delta/dt^2 \qquad \qquad .....(8.4)$$

angular acceleration of the motor

$$\alpha = d^2\theta/dt^2 = d^2\delta/dt^2 \text{ electrical radian/s}^2 \qquad \qquad .....(8.5)$$

Now, Let us consider,

$\qquad T_e$ = electromagnetic torque and

$\qquad T_s$ = shaft torque.

Neglecting the losses, the accelerating or decelerating torque may be expressed as the difference between the shaft torque and the electromagnetic torque.

$$T_a = T_s - T_e \qquad \qquad .....(8.6)$$

T_a presents accelerating torque for positive value of T_a. T_a presents decelerating torque for negative value of T_a.

Let $\quad \omega$ = Synchronous Speed of the motor.

$\qquad$ J = moment of inertia of the motor

$\qquad$ M = angular moment of the motor

$\qquad P_s$ = Mechanical Power Input

$\qquad P_e$ = electrical Power input

$\qquad P_a$ = Accelerating power input

Now M = JW $\qquad \qquad .....(8.7)$

Multiplying both sides of eq.(8.6) by ω we get

$\omega T_a = \omega T_s - \omega T_e$

$P_a = P_s - P_e$

But $J\, d^2\theta/dt^2 = T_a$

$J\, d^2\delta/dt^2 = T_a$

$\omega J\, d^2\delta/dt^2 = \omega T_a$

$\omega J\, d^2\delta/dt^2 = \omega T_a$

$M\, d^2\,\delta/dt^2\, P_a = P_s - P_e$ .....(8.8)

From eq.(8.8) gives the relation between the accelerating power and angular acceleration. It is named as swing equation. It is named as second order non linear differential equation. As the swing equation represents swing in the power angle δ during transient, we can describe stability with this differential equation.

8.8 Swing Curves

The solution of swing equation gives the variation of δ (in electrical radians) with respect to time (in seconds). The graph when plotted is called as a swing curve. It provides information regarding stability. If δ increases continuously with time, the system is unstable. While if δ starts decreasing after reaching a maximum value, it is inferred that the system will remain stable.

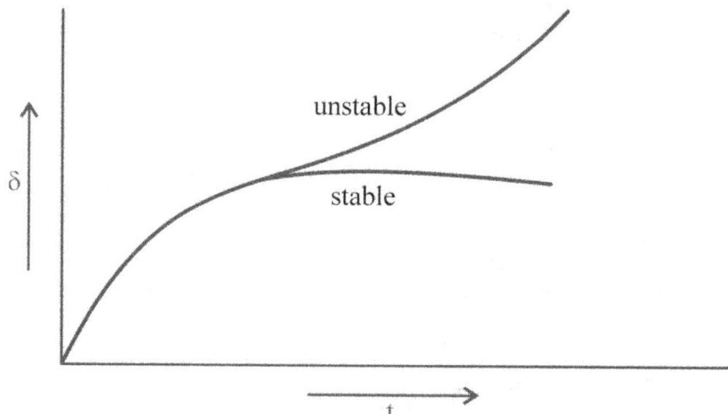

Fig.8.3 Swing Curves.

8.8.1 Use of Swing Curves

Swing curves are useful in determining the adequacy of relay protection on power systems with regard to the clearing of faults, before one or more machines become unstable and fall out of synchronism. The critical clearing time can be determined to specify the correct speed of the circuit breaker to make the system stable.

To plot the swing curves, the swing equation is to be solved. It can be solved by step by step (or point by point) method, Euler's method and Runga-Kutta method. At present, digital computer is used for solving swing equation.

8.9 Constants used in Stability Analysis

8.9.1 Inertia Constant M

From the swing equation

$$M \frac{d^2\delta}{dt^2} = P_a \qquad \qquad .....(8.9)$$

If power is in watts, δ in radians and t in sec, then M is in watts/radian/sec^2 or watt sec^2/radian. Since 1 joule = 1 watt sec, M can have its unit as joule sec/radian. If power is in MW, then M has MJ sec/radian.

If δ is specified in electrical degrees, then M has unit as MJ-sec/electrical degree. If power is in pu, then M has pu power sec^2/electrical degree. The value of M will be referred to per unit value of M.

In general, M constant may be defined as the power in MW required to produce unit angular acceleration.

8.9.2 Kinetic Energy N

If the kinetic energy of the rotor at synchronous speed is denoted by N, it is given by,

$$N = \frac{1}{2} M\omega_s \qquad \qquad .....(8.10)$$

where
$$\omega_s = 2\pi n_s; \text{ mechanical radian/sec}$$
$$= 2\pi f \text{ electrical radian/sec}$$
$$= 360 \, f \text{ electrical degree/sec}$$

where n_s is the speed in rps and f is the frequency in Hz.

Normally, generators of same MVA ratings may have different values of kinetic energy and momentum. To express them in a common way, we use a constant H (also called as Inertia constant)

8.9.3 Inertia Constant H

It is defined as the ratio of stored kinetic energy to volt ampere rating of machine.

$$H = \frac{\text{Kinetic energy}}{\text{MVA rating}} \text{ MJ/MVA}$$

$$= \frac{N}{S} \text{ where S is the MVA rating}$$

$$\therefore \, N = SH \qquad \qquad .....(8.11)$$

8.9.4 Equivalent H Constant

Consider a system in which 'n' number of generators are connected in parallel to the same bus bar.

Let

$S_1, S_2, S_3..........S_n$ be the MVA ratings of individual machines

$H_1, H_2, H_3...........H_n$ be the inertia constants of individual machines

$N_1, N_2, N_3.........N_n$ be the kinetic energy stored in individual machines

S_e be the MVA ratings of equivalent machine

H_e be the inertia constants of equivalent machine

N_e be the kinetic energy stored in equivalent machine and

S_b be the base MVA.

Energy stored by the equivalent machine is given by the sum of energies stored by individual machines.

$$N_e = N1 + N2 + ...+N_n$$

$$S_e H_e = S_1 H_1 + S_2 H_2 +.........+S_n H_n$$

where

$$S_e = S_1 + S_2 +............+ S_n$$

If the base MVA S_b is equal to the combined MVA rating of individual machines S_e

i.e., $S_b = S_e$ we get $H_e = H_1 \left(\dfrac{S_1}{S_b} \right) + H_2 \left(\dfrac{S_2}{S_b} \right) + ... + H_n \left(\dfrac{S_n}{S_b} \right)$

If the machines are identical,

We have

$$S_1 = S_2 = ...S_n = S$$

$$H_1 = H_2 = ...H_n = H$$

Then

$$H_e = n \times \frac{HS}{S_b}$$

With identical machines

$$S_b = S_e = n \times S$$

Upon substituting S_b, we get

$$H_e = n \times \left(\frac{HS}{nS} \right)$$

$H_e = H$

Thus the equivalent H constant of several identical machines operating in parallel is the same as that of any one of the machines.

8.9.5 Equivalent M Constant of Two Machines

Two synchronous machines connected by a reactance can be replaced by one equivalent machine connected through a reactance to an infinite bus as follows.

The swing equation of machine 1 is given by

$$M_1 \frac{d^2 \delta_1}{dt^2} = P_{i1} - P_{u1} \qquad \qquad .....(8.12)$$

The swing equation of machine 2 is given by

$$M_2 \frac{d^2 \delta_2}{dt^2} = P_{i2} - P_{u2} \qquad \qquad .....(8.13)$$

From eq. (8.12)

$$\frac{d^2 \delta_1}{dt^2} = \frac{P_{i1} - P_{u1}}{M} \qquad \qquad .....(8.14)$$

From eq. (8.13)

$$\frac{d^2 \delta_2}{dt^2} = \frac{P_{i2} - P_{u2}}{M_2} \qquad \qquad .....(8.15)$$

Subtracting eq. (8.15) and (8.14) we get

$$\frac{d^2 \delta_1}{dt^2} - \frac{d^2 \delta_2}{dt^2} = \frac{P_{i1} - P_{u2}}{M_2} - \frac{P_{i2} - P_{u1}}{M_1}$$

We can write

$$\frac{d^2 (\delta_1 - \delta_2)}{dt^2} = \frac{M_2 (P_{i1} - P_{u1}) - M_1 (P_{i2} - P_{u2})}{M_1 M_2} \qquad \qquad .....(8.16)$$

If δ is the relative angle between the rotors of two machines, then

$\delta = \delta_1 - \delta_2$

We can write eq. (8.16) as

$$\frac{d^2\delta}{dt^2} = \frac{M_2 P_{i1} - M_1 P_{i2}}{M_1 M_2} - \frac{M_2 P_{u1} - M_1 P_{u2}}{M_1 M_2} \qquad .....(8.17)$$

Multiply on both sides of the above equation by

$$\frac{M_1 M_2}{M_1 + M_2}$$

We get

$$\frac{M_1 M_2}{M_1 + M_2} \frac{d^2\delta}{dt^2} = \frac{M_1 M_2}{M_1 + M_2} \left(\frac{M_2 P_{i1} - M_1 P_{i2}}{M_1 M_2} - \frac{M_2 P_{u1} - M_1 P_{u2}}{M_1 M_2} \right)$$

$$\frac{M_1 M_2}{M_1 + M_2} \frac{d^2\delta}{dt^2} = \frac{M_2 P_{i1} - M_1 P_{i2}}{M_1 + M_2} - \frac{M_2 P_{u1} - M_1 P_{u2}}{M_1 + M_2} \qquad .....(8.18)$$

The swing equation of an equivalent machine is given by

$$M' \frac{d^2\delta}{dt^2} = P_i' - P_u' \qquad .....(8.19)$$

From which we can say that the equivalent values of M' P_i' and P_u' are given by

$$M' = \frac{M_1 M_2}{M_1 + M_2}$$

$$P_i' = \frac{M_2 P_{i1} - M_1 P_{i2}}{M_1 + M_2} \quad \text{and}$$

$$P_u' = \frac{M_2 P_{u1} - M_1 P_{u2}}{M_1 + M_2}$$

8.9.6   Equal Area Criterion of Stability

The transient stability of a one-machine infinite bus system or an equivalent two machine system can be assessed using equal area criterion. The expression for equal area criterion can be derived from the swing equation. Thus, by employing equal area criterion approach to determine the stability of the system, the swing equation need not be solved.

The present stability criterion may be used to evaluate the extra amount of mechanical power which can be applied at stable operating zone for a power system for the synchronous generator. Let us assume a generator is supplying power to infinite bus as shown in Figure 8.4. The initial input mechanical power, $Ps0$ and corresponding output electrical operating point is 'a(P_{s0}, ∂_0) on line power angle curve. Now after sudden change of input mechanical power, $Ps1$ and corresponding output electrical operating point is 'b(Ps_1, ∂_1)' on line power angle curve and rotor speed more than synchronous speed. Due to extra ($Ps_1 - Ps_0$) amount of power, the rotor will accelerate and load angle increased. After point 'b', rotor retards because of power transfer to the grid is more than input power Ps1. As consequences   rotor will decelerates until point 'c', again rotor gain synchronous speed.

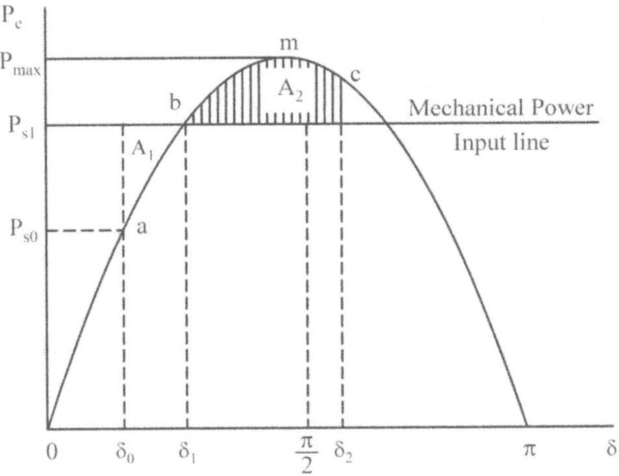

Fig.8.4

Consider the swing equation,

$$M\frac{d^2\delta}{dt^2} = P_a$$

Or,

$$M\frac{d^2\delta}{dt^2} = P_s - P_e$$

Now, multiplying both sides by 2dδ/dt, we get

$$\left(2\frac{d\delta}{dt}\right)\left(M\frac{d^2\delta}{dt^2}\right) = \left(2\frac{d\delta}{dt}\right)(P_s - P_e)$$

$$2M\frac{d^2\delta}{dt^2}\frac{d\delta}{dt} = 2(P_s - P_e)\frac{d\delta}{dt}$$

$$\left(2\frac{d^2\delta}{dt^2}\frac{d\delta}{dt}\right) = \frac{2}{M}(P_s - P_e)\frac{d\delta}{dt}$$

Integrating both sides w.r.t time, we get,

$$2\int\frac{d^2\delta}{dt^2}\frac{d\delta}{dt}dt = \frac{2}{M}\int(P_s - P_e)\frac{d\delta}{dt}dt$$

Or,

$$\frac{2}{2}\left(\frac{d\delta}{dt}\right)^2 = \frac{2}{M}\int (P_s - P_e)\,d\delta$$

Or,

$$\left(\frac{d\delta}{dt}\right)^2 = \frac{2}{M}\int (P_s - P_e)\,d\delta \qquad\qquad .....(8.20)$$

Now, for synchronous operation $\dfrac{d\delta}{dt} = 0$ i.e., the rate of change of load angle should be zero.

Thus, for synchronous running condition the limits for eq.(8.20) are in the range δ_0 and δ_2, since at $\dfrac{d\delta}{dt} = 0$, the value of δ varies from δ_0 to δ_2

Thus equation (8.20) now becomes,

$$\left(\frac{d\delta}{dt}\right)^2 = \frac{2}{M}\int_{\delta_0}^{\delta_2} (P_s - P_e)\,d\delta$$

Or,

$$\frac{d\delta}{dt} = \sqrt{\frac{2}{M}\int_{\delta_0}^{\delta_2} (P_s - P_e)\,d\delta}$$

For stability, the condition $\dfrac{d\delta}{dt} = 0$ exists,

Thus,

$$\frac{d\delta}{dt} = \sqrt{\frac{2}{M}\int_{\delta_0}^{\delta_2} (P_s - P_e)\,d\delta} = 0$$

Or,

$$\sqrt{\frac{2}{M}\int_{\delta_0}^{\delta_2} (P_s - P_e)\,d\delta} = 0$$

Or,

$$\frac{2}{M}\int_{\delta_0}^{\delta_2} (P_s - P_e)\,d\delta = 0$$

Or,

$$\int_{\delta_0}^{\delta_2} (P_s - P_e)\,d\delta = 0$$

Or,

$$\int_{\delta_0}^{\delta_1}\left(P_s - P_e\right)d\delta \neq \int_{\delta_1}^{\delta_2}\left(P_s - P_e\right)d\delta = 0$$

Or,

$$\int_{\delta_0}^{\delta_1}\left(P_s - P_e\right)d\delta = -\int_{\delta_1}^{\delta_2}\left(P_s - P_e\right)d\delta$$

Here $\int_{\delta_0}^{\delta_1}\left(P_s - P_e\right)d\delta = A_1$ designates the positive or accelerating area.

And $\int_{\delta_1}^{\delta_2}\left(P_s - P_e\right)d\delta = A_2$ designates the negative or decelerating area

Or in other words, $A_1 = -A_2$ .....(8.21)

The relationship obtained from eq.(8.21) demonstrates the equal-area criterion of stability.

For the above criterion the value of δ_1 should always be less than or equal to 90^0 (i.e., $\delta_1 \leq 90^0$), but as long as the equal-area criterion is satisfied then values for δ_2 greater than 90^0 can also be entertained.

The simplicity of equal-area criterion of stability to determine the stability of the system (in that the system is not stable if $A_2 \neq A_1$) can be taken advantage of to determine the maximum power that can be transferred within the stability limit.

For the system to remain within the stability limit, the condition $A_1 = A_2$, must be fulfilled.

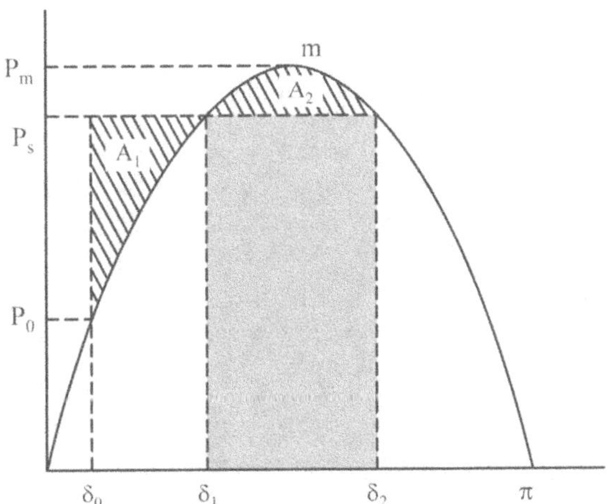

Fig.8.5 Transient Stability Limit.

Consider the given figure showing a power angle curve wherein is δ_2 is greater than 90^0

Here, $P_s = P_m \sin\delta_1 = P_m \sin \delta_2$

Where, $\delta_1 = \delta_c$ = critical torque angle

δ_2 in terms of δ_1 is, $\delta_2 = \pi - \delta_1$

For the system to be in stable condition, the condition $A_1 = A_2$ should be satisfied,

Thus,

$$\int_{\delta_0}^{\delta_1}(P_s - P_m \sin \delta)\, d\delta = \int_{\delta_1}^{\delta_2}(P_m \sin \delta - P_s)\, d\delta \qquad .....(8.22)$$

Or, $P_s(\delta_1 - \delta_0) - \int_{\delta_0}^{\delta_1} P_m \sin \delta\, d\delta = \int_{\delta_1}^{\delta_2} P_m \sin \delta\, d\delta - P_s(\delta_2 - \delta_1)$

Or, $P_s(\delta_1 - \delta_0) + P_m (\cos \delta_1 - \cos \delta_0) = P_s(\delta_1 - \delta_2) + P_m (\cos \delta_1 - \cos \delta_2)$ .....(8.23)

Putting the value of $P_s = P_m \sin \delta_1$ in the above eqn (8.23) we obtained

$$(\delta_1 - \delta_0) \times P_m \sin \delta_1 + P_m (\cos \delta_1 - \cos \delta_0)$$

$$= (\delta_1 - \delta_2) \times P_m \sin \delta_1 + P_m (\cos \delta_1 - \cos \delta_2) \qquad .....(8.24)$$

Or, $(\delta_2 - \delta_0) \times \sin \delta_1 + (\cos \delta_2 - \cos \delta_0) = 0 \qquad .....(8.25)$

The above condition can be used for finding stability for the case of Sudden change of Power.

Equal Area Criterion can also be used to analyze the stability for (i) sustained line fault and (ii) line fault cleared after some time by breaker operation.

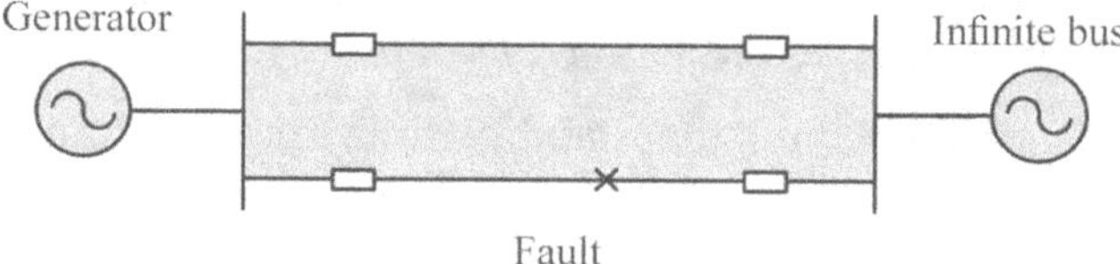

Fig.8.6 Parallel Feeder.

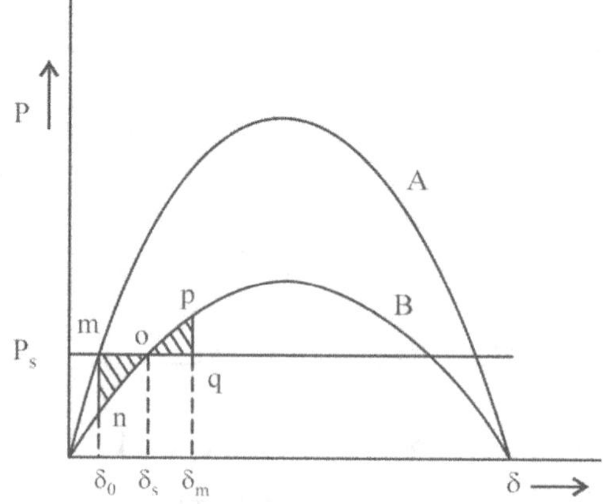

Fig.8.7 Curve A Power angle Curve under Healthy condition; Curve B – Power Angle Curve under Faulted Condition.

For condition (i)

Fig.(8.6) shows a parallel feeder connecting a generator to an infinite bus system. In Fig.(8.7), Curve A represents the power angle curve for a healthy system and Curve B shows the power angle curve for a system with line fault. Also in Fig.(8.7), the input to the generator (P_S) and the voltage behind transient reactance are assumed constant. Curve B is lower than Curve A since in the event of fault and its subsequent clearing with the operation of circuit breakers, there will be an increase in the equivalent impedance between the bus bars. Initially the generator torque angle is at δ_0 for an initial generator input of P_s but after occurrence of fault and its instantaneous clearance the generator output falls to point n on curve B, with the generator input remaining constant at P_s, resulting in the generator input greater than the generator output. As a result, the rotor accelerates and the torque angle correspondingly increases to recover to point o from point n along curve B. The acceleration stops when point o is reached but the speed of the generator continues to increase. From point o to p the rotor tries to decelerate till a certain point p is reached where the relative speed becomes equal to zero and the torque angle stops increasing. At point p the situation is now reversed in that the generator output is now more than the input, and the cycle repeats in reverse (i.e. deceleration from point p to point o, then overshoots point o and goes towards point n due to the continued decrease of the torque angle). This cycle continues to repeats itself if damping is not achieved. In actual practice due to presence of damping in the rotor, it operates at point o on curve B with a torque angle of δ_s. For such cases, in order to determine the transient stability limit, the input line P_s should be raised such that equal area distribution is achieved for the shaded regions illustrated in Fig.8.8. The value of P_s (in this case) is known as the transient stability limit.

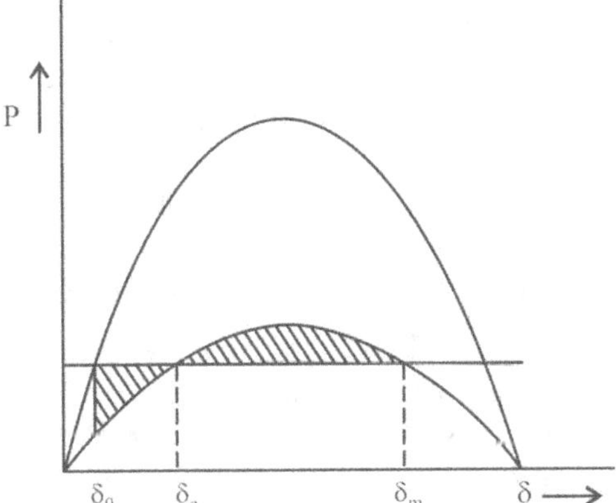

Fig.8.8 Transient Stability Limit for System in Fig.8.6.

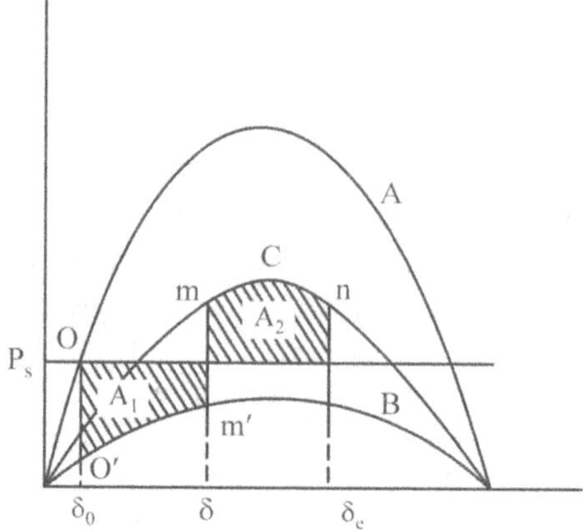

Fig.8.9 Equal Area Criterion applied to system of Fig.8.6,
for a fault cleared at an angle $\delta < \delta_0$.

For condition (ii)

For this condition as well, the parallel feeder illustrated in Fig.(8.6) is considered. The power angle curves for study of the equal area criterion is illustrated in Fig.(8.9). Similar to the previous case, here too the Curve A represents the power angle curve in healthy condition whereas the Curve B represents the power angle curve for the system with line fault at one of the lines, and the Curve C represents the power angle curve with the faulty line removed.

At the initial state the torque angle is considered δ_0 for generator input of P_s. At the instant of fault, the output of the generator drops to point O' of curve B. Now, since the generator input remains constant, automatically the input becomes greater than the output and thus the rotor accelerates and moves up along the curve B to point m' (since the fault is removed after some time has elapsed). When the fault is removed after some time, the operating point now shifts to point m on the Curve C. At this point, the output is more than the input and so the speed of the rotor reduces till the speed of the generator and the speed of the infinite bus are matched and the torque angle fluctuation ceases at point n.

From the illustration in Fig.(8.9), the relationship between transient stability limit, the type of disturbance and the clearing time of the breaker can be clearly observed. It can be observed that the faster the operations of the breaker the smaller will the area A_1 be resulting in a larger transient stability limit.

8.10 Critical Clearing Angle

From the previous discussions it can be observed that for any given initial load condition there will always be an associated critical clearing angle. The system tends to be unstable if the clearing angle is greater than the critical value and vice-versa.

To determine the value of critical clearing for a given initial load,

Let the three power angle curves (Fig.8.10) be represented as

$A = P_m \sin \delta$ before the fault

$B = r_1 P_m \sin \delta$ during the fault

$C = r_2 P_m \sin \delta$ after the fault

For transient stability limit, the two areas $A_1 = A_2$ or equivalently the area under the curve abcd (a rectangle) should be equal to the area under the curve dfghbc i.e.,

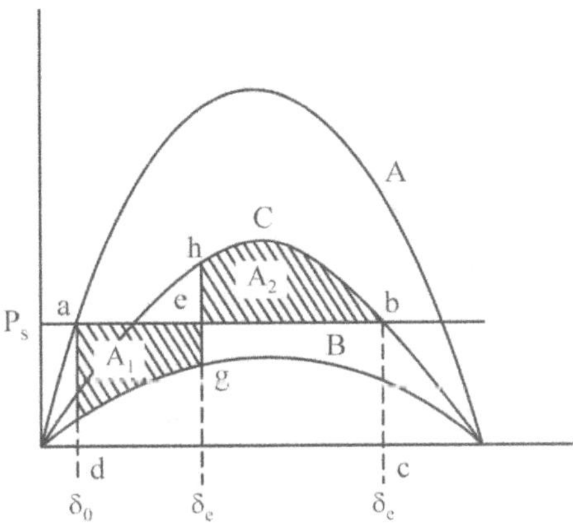

Fig.8.10 Critical Clearing Angle.

$$(\delta_c - \delta_0)P_s - \int_{\delta_0}^{\delta_c} r_1 P_m \sin \delta \, d\delta = \int_{\delta_c}^{\delta_m} r_2 P_m \sin \delta \, d\delta - (\delta_m - \delta_c)P_s$$

Or, $(\delta_c - \delta_0) P_s - r_1 P_m[\cos \delta_0 - \cos \delta_c] = r_2 P_m[\cos \delta_c - \cos \delta_m] - (\delta_m - \delta_c) P_s$

Or, $(\delta_m - \delta_0) P_s = r_1 P_m[\cos \delta_0 - \cos \delta_c] + r_2 P_m[\cos \delta_c - \cos \delta_m]$

Now substituting $P_s = P_m \sin \delta_0$

$$(\delta_m - \delta_0)P_m \sin \delta_0 = r_1 P_m[\cos \delta_0 - \cos \delta_c] + r_2 P_m[\cos \delta_c - \cos \delta_m]$$

$$(\delta_m - \delta_0)\sin \delta_0 = (r_2 - r_1) \cos \delta_c + r_1 \cos \delta_0 - r_2 \cos \delta_m$$

$$\therefore \quad \cos \delta_c = \frac{(\delta_m - \delta_0) \sin \delta_0 - r_1 \cos \delta_0 + r_2 \cos \delta_m}{r_2 - r_1} \qquad .....(8.26)$$

Now from the curves,

$$P_s = P_m \sin \delta_0 = r_2 P_m \sin \delta_m = r_2 P_m \sin(\pi - \delta_m)$$

or $\sin \delta_0 = r_2 \sin(\pi - \delta_m)$

or $\delta_m = \pi - \sin^{-1}\left(\dfrac{\sin \delta_0}{r_2}\right)$

Thus if r_1, r_2 and δ_0 are known, the critical clearing angle δ_c, can be obtained.

Problem 8.2: A lossless alternator supplies 39.5 MW to an infinite bus bar. The steady state stability limit is 79 MW. Determine if the alternator will remain stable if the input of the prime mover abruptly increased by 31.6 MW.

Solution:

Let, the shaft power be given by P_s.

The electrical power be given by P_e.

The steady state stability limit $= P_{max} = 79$ MW

The power angle curve can be drawn as:

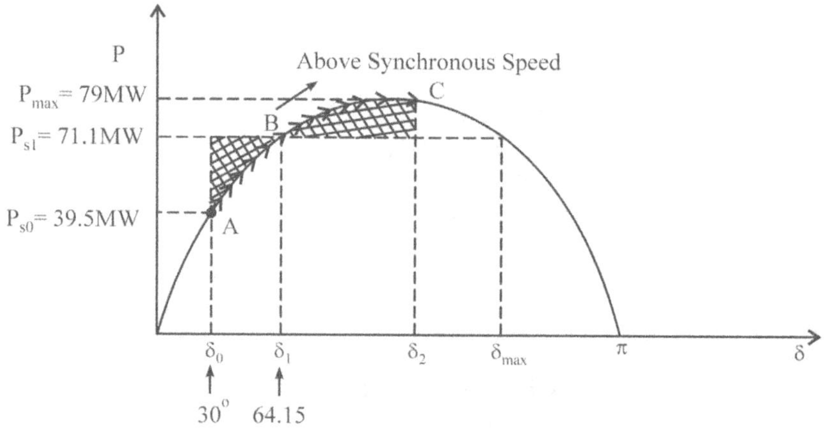

Fig.8.11 Power Angle Curve.

When the alternator is supplying 39.5 MW:

Let the angle at 39.5 MW supply on power angle curve be δ_0.

As,

$P_s = P_e = P_{max} \sin \delta$

Then,

$P_{s0} = P_{max} \sin \delta_0$

$39.5 = 79 \sin \delta_0$

$\sin \delta_0 = 1/2 = 0.5$

So, $\delta_0 = 30° = 0.5235$ radian

When the alternator is supplying 71.1 MW:

Let the angle at 71.1 MW supply on power angle curve be δ_1.

Then,

$P_{s1} = P_{max} \sin \delta_1$

$71.1 = 79 \sin \delta_1$

$\sin \delta_1 = 71.1/79 = 0.9$

So, $\delta_1 = 64.15° = 1.0725$ radian

The alternator will become unstable if:

$\delta_2 > \delta_{max}$ .... (unstable)

The alternator will be stable if:

$\delta_2 < \delta_{max}$.... (stable)

Calculating δ_{max}:

$\delta_{max} = \pi - \delta_1 = 115.84°$

Applying equal area criteria:

$$\int_{\delta_0}^{\delta_1}\left(71.1 - 79\sin\delta\right)d\delta = \int_{\delta_1}^{\delta_2}\left(79\sin\delta - 71.1\right)d\delta$$

$79 \cos \delta_2 + 71.11\, \delta_2 - 105.6438 = 0$

Solving by numerical method, we get:

$\delta_2 = 40.32° = 105.6257$ radian

Hence,

$\delta_2 > \delta_{max}$

So, the alternator will be in stable condition.

Problem 8.3: A synchronous generator, capable of developing 83MW power per phase, operates at a power angle of 8°. By how much can the input shaft power be increased suddenly without loss of stability? Assume that P(max) will remain constant.

Solution:

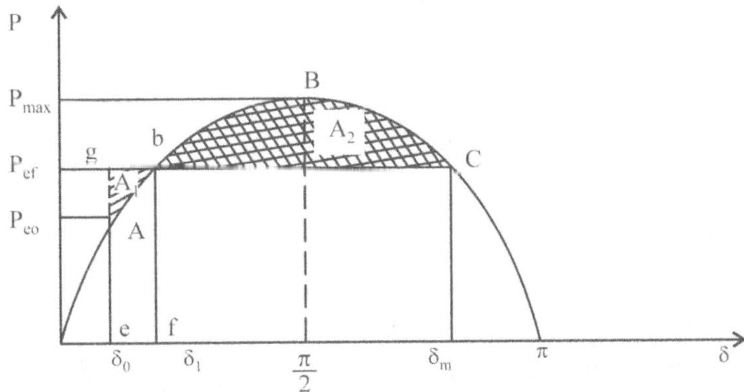

Initially, $\delta_0 = 8°$

$P_{e0} = P_{(max)} \sin\delta_0 = 83 \sin 8° = 11.55$MW

Let δ_m be the power angle to which the rotor can swing before losing synchronism. If this angle is exceeded, Pi will again become greater than Pe and the rotor will once again be accelerated and synchronism will be lost as shown in the figure.

Therefore, according to equal area criterion-

$A_1 = A_2$

Area agb = area bBc

Area of gbfe – area of abfe =area of bBchf – area bchf

$$P_{ef}(\delta_1 - \delta_0) - \int_{\delta_0}^{\delta_1} P_{(max)} \sin\delta d\delta = \int_{\delta_1}^{\delta_m} P_{(max)} \sin\delta d\delta - P_{ef}(\delta_m - \delta_1)$$

Since $P_{ef} = P_{(max)} \sin\delta_0$, put in above equation

$$P_{(max)} \sin\delta_0(\delta_1 - \delta_0) - \int_{\delta_0}^{\delta_1} P_{(max)} \sin\delta d\delta = \int_{\delta_1}^{\delta_m} P_{(max)} \sin\delta d\delta - P_{(max)} \sin\delta_0 (\delta_m - \delta_1)$$

Take common P(max) from both side:

$$\sin\delta_0(\delta_1 - \delta_0) - \int_{\delta_0}^{\delta_1} \sin\delta d\delta = \int_{\delta_1}^{\delta_m} \sin\delta d\delta - \sin\delta_0 \left(\delta_m - \delta_1\right)$$

$$\sin\delta_0(\delta_1 - \delta_0) - (\cos\delta_0 - \cos\delta_1) = \cos\delta_1 - \cos\delta_m - \sin\delta_0(\delta_m - \delta_1)$$

since $\delta_m = \pi - \delta_1$ put in above equation

$$\sin\delta_0(\delta_1 - \delta_0) - (\cos\delta_0 - \cos\delta_1) = \cos\delta_1 - \cos(\pi - \delta_1) - \sin\delta_0(\pi - \delta_1 - \delta_1)$$

$$\sin\delta_0(\delta_1 - \delta_0) - (\cos\delta_0 - \cos\delta_1) = \cos\delta_1 + \cos\delta_1 - \sin\delta_0(\pi - 2\delta_1)$$

put $\delta_0 = 8° = 0.139$ radian

$$\sin 8° (\delta_1 - 0.139) - (\cos 8° - \cos\delta_1) = 2\cos\delta_1 - \sin 8°(3.14 - 2\delta_1)$$

$$0.139(\delta_1 - 0.139) - (0.99 - \cos\delta_1) = 2\cos\delta_1 - 0.139(3.14 - 2\delta_1)$$

$$\delta_1(0.139 - 2 \times 0.139) - 0.139 \times 0.139 - 0.99 + 0.139 \times 3.14 = -\cos\delta_1$$

$$-0.139\delta_1 - 0.573 = -\cos\delta_1$$

$$\cos\delta_1 - 0.139 \delta_1 = 0.573$$

after solving, we get $\delta_1 = 65.5°$

Now, $P_{ef} = P_m \sin\delta_1 = 83\sin 65.5°$

$$= 75.52 MW$$

Initial power developed by machine was 11.55MW. Hence without loss of stability, the system can accommodate a sudden increase of-

$$\mathbf{P_{ef} - P_{e0}} = 75.52 - 11.55 = 63.97 \text{ MW per phase}$$

$$= 3 \times 63.97$$

$$= 191.91 \text{ MW } (3 - \varphi) \text{ of input shaft power.}$$

Problem 8.4: A loss free generator supplies 75 MW to an infinite bus, the steady state limit of the system being 150 MW. Determine whether the generator will remain in synchronism if the prime mover input is abruptly increased by 45 MW.

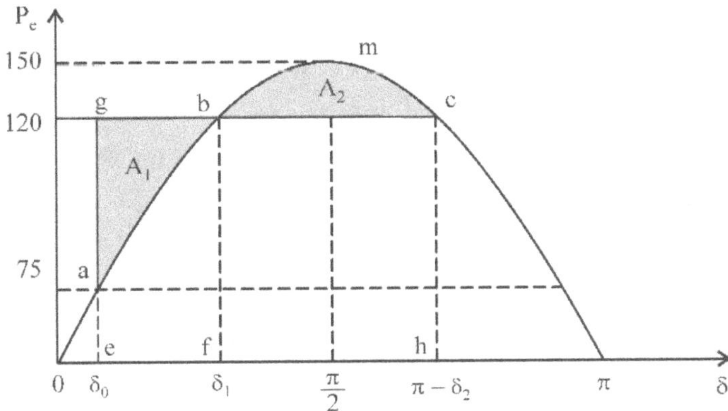

Solution:

The power- angle curve is shown in fig. The equation of the power-angle curve is

$$P_e = P_{max} \sin \delta \qquad \qquad .....(i)$$

The initial operating point is at a where $P_e = 50$ MW. The prime mover input is abruptly increased by 45 MW. The desired operating point is at 'b' so that $ag = 30$ MW. The maximum swing of the rotor can be up to point c. The system will be stable if the area agb is less than area bmc. The load angles corresponding to point a and b can be found in eq. (i).

For point a, $P_e = 75$, $\delta = \delta_1$

Substituting these values in eq. (i) we get

$$75 = 150 \sin \delta_1$$

$$\sin\delta_1 = \frac{75}{150}, \quad \delta_1 = 30° = 0.523 \text{ rad}$$

For point b, $P_e = 120°$ $\delta = \delta_2$

$$120 = 150 \sin\delta_2$$

$$\sin\delta_2 = \frac{120}{150}, \quad \delta_2 = 53.13° = 0.927 \text{ rad}$$

A_1 = area agb = area of rectangle **egbf** – area **eabf** under the power angle sine curve

$$= (eg \times ef) - \int_{\delta_0}^{\delta_1} P_e d\delta - 120(\delta_1 - \delta_0) - \int_{\delta_0}^{\delta_1} 150 \sin \delta \, d\delta$$

$$= 120 (0.927 - 0.523) + 150 (\cos \delta_1 - \cos\delta_0)$$

$$= 48.48 + 150(0.600 - 0.866)$$

$$= 48.48 - 39.9$$

$= 8.1$ MW rad

$A_2 =$ area **bmc** = area **bmchf** under the power angle sine curve – area of rectangle **bchf**

$$= \int_{\delta_1}^{\pi-\delta_1} P_e d\delta - 120(\pi - \delta_1 - \delta_1) = \int_{\delta_1}^{\pi-\delta_1} 150\, sin\ \delta\, d\delta - 120(\pi - 2\delta_1)$$

$= -150[\cos(\pi - \delta_1) - \cos \delta 1] - 120(\pi - 2 \times 0.927)$

$= 300 \cos\delta_1 - 154.32 = 300 \times 0.6 - 154.32 = 180 - 154.3 = 25.68$ MW rad

Since area acb (A_1) is less than area bmf (A_2) the system is stable.

Exercise

1. What do you understand by the term steady state stability? Explain briefly.

2. What is difference between transient and steady state stability?

3. On what basis do you conclude that a given synchronous machine has lost stability?

4. Explain the swing equation of a single machine.

5. What do you mean by equal area criteria?

6. What is equal area criterion? Discuss its application and limitation in the study of power system stability.

7. Differentiate between steady state stability and transient stability of power system. Discuss the factors effecting steady state stability and transient stability of power system

8. Derive an expression for Power Angle curve under steady state operation while a synchronous generator is connected to an infinite bus.

Multiple Choice Question

1. Consider a lossy transmission line with V_1 and V_2 as the sending and receiving end voltages, respectively. Z and X are the series impedance and reactance of the line, respectively. The steady-state stability limit for the transmission line will be

    (A) greater than $\dfrac{V_1 V_2}{X}$ (B) less than $\dfrac{V_1 V_2}{X}$

    (C) equal to $\dfrac{V_1 V_2}{X}$ (D) equal to $\dfrac{V_1 V_2}{Z}$

    (GATE)

2. The angle δ in the swing equation of a synchronous generator is the

    (A) angle between stator voltage and current

    (B) angular displacement of the rotor with respect to the stator.

    (C) angular displacement of the stator mmf with respect to a synchronously rotating axis.

    (D) angular displacement of an axis fixed to the rotor with respect to a synchronously rotating axis.

(GATE)

3. Steady state stability of a power system is the ability of the power system to

   (A) Maintain voltage at the rated voltage level.

   (B) Maintain frequency exactly at 50 Hz.

   (C) Maintain a spinning reserve margin at all times.

   (D) Maintain synchronism between machines and on external tie lines.

(GATE)

4. A 100 MVA, 11 Kv, 3-phase, 50 Hz, 8-pole synchronous generator has an inertia constant H equal to 4 MJ/MVA. The stored energy in the rotor of the generator at synchronous speed will be $H = \dfrac{E}{G}$

   (A)   100 MJ   (B) 400 MJ

   (C)   800 MJ   (D) 12.5 MJ

(GATE)

5. During a disturbance on synchronous machine, the rotor swings from A to B before finally settling down to a steady state at a point C on the power angle curve. The speed of the machine during oscillation is synchronous at point(s)

   (A) A and B   (B) A and C

   (C) B and C   (D) Only at C

(GATE)

6. The transient stability of the power system can be effectively improved by
   (A) Excitation control
   (B) Phase shifting transformer
   (C) Single pole switching of circuit breakers
   (D) Increasing the turbine valve opening

(GATE)

7. A round rotor generator with internal voltage $E_1 = 2.0$ p.u. and $X = 1.1$ p.u. is connected to a round rotor synchronous motor with internal voltage $E_2 = 1.3$ p.u. and $X = 1.2$ p.u. The reactance of the line connecting the generator to the motor is 0.5 p.u. when the generator supplies 0.5 p.u. power, the rotor angle difference between the machines will be

   (A) 57.42° (B) 1°
   (C) 32.58° (D) 122.58^0

(GATE)